NUCLEAR ENERGY

Nuclear Energy

Boom, Bust, and Emerging Renaissance

Edward A. Friedman

OXFORD
UNIVERSITY PRESS

2 04

OXFORD
UNIVERSITY PRESS

Great Clarendon Street, Oxford, OX2 6DP,
United Kingdom

Oxford University Press is a department of the University of Oxford.
It furthers the University's objective of excellence in research, scholarship,
and education by publishing worldwide. Oxford is a registered trade mark of
Oxford University Press in the UK and in certain other countries.

Published in the United States of America by Oxford University Press
198 Madison Avenue, New York, NY 10016, United States of America

British Library Cataloguing in Publication Data
Data available

Library of Congress Control Number: 2025934184

ISBN 9780198925798
ISBN 9780198925781 (pbk.)

DOI: 10.1093/9780198925811.001.0001

Printed and bound by
CPI Group (UK) Ltd., Croydon, CR0 4YY

Links to third-party websites are provided by Oxford in good faith
and for information only. Oxford disclaims any responsibility for the materials
contained in any third-party website referenced in this work.

The manufacturer's authorised representative in the EU for product safety is
Oxford University Press España S.A., Parque Empresarial San Fernando de
Henares, Avenida de Castilla, 2 – 28830 Madrid (www.oup.es/en or
product.safety@oup.com). OUP España S.A. also acts as importer into Spain of
products made by the manufacturer.

Acknowledgements

I thank the Stevens Institute of Technology, and in particular Dean Kelland Thomas, for the support and encouragement I received for teaching a course on nuclear energy policy that provided the foundation for this book.

I was able to devote the time and attention necessary to develop this book and to enjoy welcome encouragement for its pursuit, thanks to my loving spouse, Dr. Arline J. Lederman.

I have been fortunate to receive good advice and support in this endeavor from my sons Dr. Millard Timur Friedman and Dr. Philip Kerim Friedman.

Conversations with my Stevens colleague, Professor Alex Wellerstein, have enhanced my perspective on this subject and deserve my appreciation. I also extend my deep thanks to Professor Wellerstein for having orchestrated the graphics and images for this book.

I have also profited from interactions with author Nick Taylor, who provided a better understanding of what constitutes a good book.

Significant contributions to this work were provided by Dylan Moon, who engaged in editorial assistance and critical advice regarding its content.

I also thank Sudheekshitha Malepati for her thoughtful attention to helping organize the reference materials for this book.

Contents

List of Figures

Introduction

"Often, the very fact that the words of science are the same as those of our common life and tongue can be more misleading than enlightening, more frustrating to understanding than recognizably technical jargon. For the words of science—relativity, if you will, or atom, or mutation, or action—have been given a refinement, a precision, and in the end a wholly altered meaning."

This observation by J. Robert Oppenheimer in his 1953 lecture series "Science and the Common Understanding" identified a major challenge for anyone wishing to communicate insights acquired in the world of science and technology to those residing in the general population. When writing this book, I have been grappling with the question of what words should be included in this narrative and the deeper question regarding the level of detail needed in their presentation. I first encountered these words of Oppenheimer as a high-school senior listening to his lectures that were broadcast on the radio in the New York City metropolitan area.

The voice of Oppenheimer, delivering those lectures, reinforced my desire to study physics and to engage in discourse in the larger society about the content of science and technology. I went on to pursue such study at MIT and at Columbia University and to engage in communications about science as a Professor of Physics at Stevens Institute of Technology.

I first encountered questions about technology's impact on civil society in the early 1960s. At that time, I put myself forward as a subject matter expert who could provide perspective on public policy questions that were being debated locally and nationally. These included the proposals for nuclear reactors at Ravenswood and Shoreham and John F. Kennedy's call for citizens to build fallout shelters in their homes.

Now, 50 years later, I find myself again searching for the most appropriate words to communicate facts about nuclear reactors and their role in society.

Together with the choice of appropriate words about nuclear issues is the question of the amount of detail to include that will allow a reader untrained in science or engineering to make informed decisions. Citizens today cannot escape from the choices facing a society threatened by the potentially devastating impact of global warming. My goal in writing this book is to empower participation by a broad segment of the public in this decision making.

The task of telling the story of nuclear energy today is quite different from what it was in the 1960s. During the intervening years we have seen considerable use of nuclear energy along with accidents that have had a chilling effect on that use. Most notably, there have been innovations that promise the realization of remarkably new, efficient, and safe reactors that are now on the horizon.

But we must remember that this is a technology that has long timelines. It takes many years for a new reactor design to move from conceptualization to implementation and once it becomes operative it is likely to be active for many years. Reactors are built to last 40, 60, and 80 years. A consequence is that of the approximately 400 nuclear reactors that are operating in the world today, a considerable majority of them were designed and built in the 1960s and 1970s.

Since the realities of nuclear energy today are deeply rooted in the past, I have chosen to present this subject from an historical perspective. Chapter 1 starts with the discovery of radioactivity, which led English chemist Frederick Soddy to predict the possibility of nuclear energy and share these insights with his friend H. G. Wells who, in 1914, wrote *The World Set Free*, in which he describes nuclear reactors and nuclear weapons. Chapter 1 concludes with information about the discovery of nuclear fission by Otto Hahn and Lisa Meitner.

Chapter 2 then discusses the development of the first reactor and the first nuclear weapons in the remarkable Manhattan Project of the Second World War. Chapter 3 deals with the post-war years when Admiral Hyman Rickover oversaw the development of a nuclear-powered submarine fleet and President Dwight D. Eisenhower introduced the Atoms for Peace program. Chapter 3 concludes with the transfer of nuclear technology to the private sector with the first civilian nuclear reactors implemented by Westinghouse and General Electric.

Subsequent chapters tell the story of the nuclear boom that began in 1960 and continued until the disaster at Chernobyl in 1986. This period of growth of nuclear energy is divided into multiple chapters: the US account in Chapter 4, the UK account in Chapter 6, France in Chapter 7, Russia in Chapter 8, China in Chapter 9, and other countries in Chapter 10.

Embedded in the boom section is Chapter 5, which deals with resistance to nuclear developments. Since widespread antipathy, fear, and deep psychological resistance to nuclear artifacts permeated society from the time of the Hiroshima detonation, it seemed best to discuss this foreboding early on in the book.

The bust part of the book is addressed in Chapters 11, 12, and 13, which cover the accidents at Three Mile Island, Chernobyl, and Fukushima. While there are other accidents of interest, these are the most consequential in influencing public attitudes. The important Windscale accident is dealt with in Chapter 6, and the Kyshtym disaster in Russia is included in Chapter 12.

The point of view of this book is that it is vital that carbon emissions be reduced in the most expeditious manner and that that will necessarily include power generated by wind, solar, and nuclear methods. To help the reader evaluate these choices, a brief review of wind and solar power and batteries is included in Chapter 14

The accidents, particularly the Chernobyl event, precipitated a hiatus in reactor construction worldwide. During this hiatus, the need for fail-safe reactors was brought into focus, resulting in designs for Generation-III and Generation-IV reactors and giving rise to the beginnings of a nuclear renaissance.

The renaissance section of the book starts with Chapter 15, which reviews Generation-III reactors. These reactors have various improved features compared with earlier models, with the key characteristic that in the first 72 hours following an accident the system avoids serious consequences without activation of technology or human intervention. The design of these systems employs the law of physics to control accidents.

Chapter 16 then reviews three distinct topics—fuel, waste, and radiation—and does so in the context of advanced Generation-III and Generation-IV reactor designs. Generation-IV designs are introduced in Chapter 17. I identify the beginning of the renaissance period for nuclear energy as the period around 2000 when an international initiative began an organized quest for developing fail-safe reactors identified as Generation IV.

The following four chapters (18, 19, 20, and 21) describe the four most promising Generation-IV designs: molten salt reactors, liquid sodium reactors, liquid lead reactors, and high-temperature gas-cooled reactors.

Chapter 21 also discusses the use of high-temperature reactors for the direct use of heat in industrial applications, such as the processing of metals. Greenhouse gas production from this sector of the US economy accounts for more than 20% of total greenhouse emissions.

These high-temperature applications cannot easily be addressed by wind and solar generators, which have difficulty reaching the high temperatures that are needed. Hence, the use of nuclear energy for this application is essential.

A significant energy need exists in Arctic regions as well as in island locations where floating nuclear reactors can be employed. This specialized application is the subject of Chapter 22.

The dominant thrust for nuclear reactor technology currently is devoted to the production of small modular reactors that generate up to 300 MW electric and can be built in factories. This pursuit, which encompasses the development of more than forty new designs, is the subject of Chapter 23.

Chapter 24 provides an overview of the nuclear reactor export market, which is stimulating initiatives in several countries and is needed to meet the energy needs of countries that do not manufacture nuclear reactors.

Finally, Chapter 25 provides a summary of current programs to meet the world's energy needs and carbon emission challenges. These initiatives need to address not only technological issues, but also financial, regulatory, and political concerns as well. Of overriding significance is Oppenheimer's concern for the ways in which words like waste, nuclear, and radiation are perceived.

1

The Energy Stored in Atoms

Radioactivity, Fission, and the Possibility of a Bomb

My mother never lost her conviction that the new radiations, in spite of their dangerous power, would prove of inestimable benefit to humanity. She would say, "I believe that radium can do more good than harm if it is used with care and intelligence."

– Ève Curie[1]

Overview

This first chapter describes surprising observations that led to the discovery of nuclear fission. These were disruptive observations that constitute a chain of discrete events, starting in 1895, that yielded a clear picture of nuclear reality in 1938. The story of these observations is a science tale in which enquiring minds sought to understand subatomic phenomena. After the discovery of unexpected and mysterious rays, many scientists sought to bring meaning and understanding to this new reality. It is noteworthy that two key players in this tale are women, who were breaking new ground for females in a traditionally male domain. One was Marie Curie, whose work was acknowledged with two Nobel Prizes and the other was Lise Meitner, who was overlooked by the Nobel Prize committee and is probably the most seriously under-recognized female scientist in history. This chapter brings us to the point, on the eve of the Second World War, when human minds confronted the challenge of applying this new knowledge of the natural world to the creation of mechanisms and devices to make a bomb.

The vast energy stored within atoms was released for the first time with the "Trinity" test at the Alamogordo Bombing Range in the New Mexico desert on July 16, 1945. That explosion marked the dawn of the Atomic Age, but the scientific discoveries leading up to that event were set in motion by two accidental observations: in Germany in 1895 and in France in 1896.

---

[1] Ève Curie, "Madame Curie and Her Work," lecture at Vassar College, 1937.

Nuclear Energy. Edward A. Friedman, Oxford University Press. © Edward A. Friedman (2025). DOI: 10.1093/9780198925811.003.0001

The Discovery of X-Rays

Triggering the series of observations leading to the realization of atomic energy was the discovery in 1895 of X-rays. Setting the stage for that discovery were observations that clarified the nature of electricity and electrical currents in the eighteenth century. These included the experiment in 1752 by Benjamin Franklin, who flew a kite to determine that lightning was an electrical phenomenon. The production of electrical currents in laboratories was then facilitated by the invention of the battery by Alessandro Volta in 1800. During the nineteenth century, using batteries, electrical currents could be generated inside vacuum tubes.

Among those who were experimenting with such electrical currents in the nineteenth century was Wilhelm Conrad Roentgen at the University of Wurzburg in Germany. Roentgen experimented with vacuum tubes that were stimulated with high voltages and covered with cardboard to prevent the emission of light from the tube. In his laboratory was a screen that was coated with material that would glow when exposed to electrical stimulation. To his amazement, this screen emitted visible light when it was nine feet away from his covered vacuum tube. From this observation on November 8, 1895, he concluded, correctly, that the tube was emitting an invisible ray that could transmit electrical energy. He immediately began systematic studies to see if these rays could be blocked by various substances and determined that the magnitude of the transmitted radiation was reduced in proportion to the density of the intermediate material.

He named the rays X-rays to underscore their mysterious origin. Using photographic plates, he found that when the X-rays passed through objects of varying density, images were produced that revealed the inner structure of the target object. Just six weeks after his first observation, on December 22, 1895, Roentgen exposed his wife's hand to these rays (Figure 1.1). When she saw the bone structure of her hand and the outline of the ring that she was wearing in the resulting image, she exclaimed, "I have seen my death."

Roentgen published his findings and publicly presented his results in January 1896. The impact was immediate. His discovery created a sensation. The opportunity to see structure within the body stimulated many to seek X-ray images of their hands and to fear that their more private parts might be exposed for all to see! Since it was relatively easy to produce X-rays, devices became available for use in homes and public locations. The medical profession quickly claimed special expertise in interpreting images and well-known inventors such as Thomas Alva Edison and Nikola Tesla sought opportunities to explore this new phenomenon. There were early signs that X-rays were dangerous as burns and irritations of the skin became evident, but it took almost a decade before safeguards were introduced. Even then, applications (e.g., the Foot-a-Scope to view shoe fittings) were introduced and in wide use until the late 1940s.

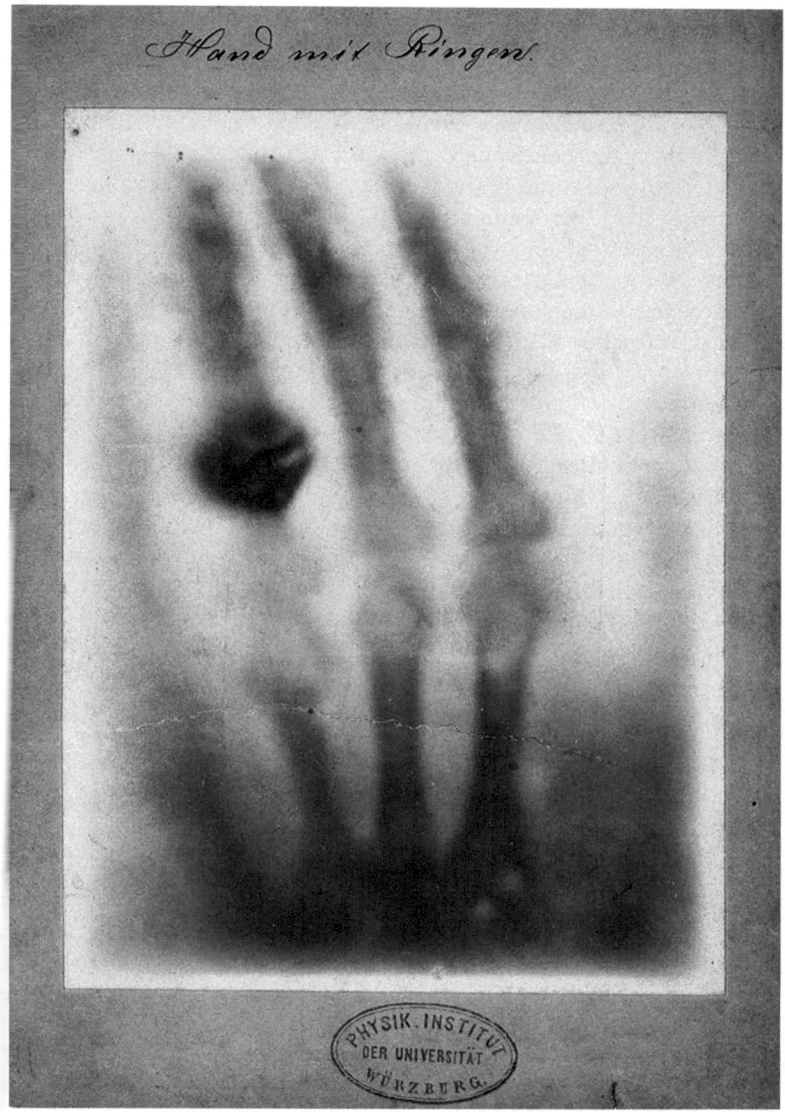

Figure 1.1 *An X-ray of Mrs. Rontgen's ringed hand (1895).*

Photoprint from radiograph by W. K. Röntgen, 1895. Wellcome Collection 32971i. https://wellcomecollection.org/works/wjc8ejn2. under a Creative Commons Attribution-NonCommercial 4.0 International (CC BY-NC 4.0).

Radioactivity

In 1896, almost immediately after Roentgen announced his observation of X-rays, Henri Becquerel, who, like Roentgen, was a physicist working in Paris, began studies of the impact of light on uranium. He exposed uranium to sunlight and then placed it in an enclosure next to a covered photographic plate. He was speculating that the uranium would absorb energy from the sunlight leading to the emission of X-rays that would penetrate the photographic plate covering thereby producing an observable mark.

At the time that Becquerel performed these experiments, uranium was a well-known element with unique chemical properties. It had been identified, more than 100 years earlier in 1789, German chemist Martin Klaproth, identified it with its characteristic mass and charge. Klaproth studied the properties of pitchblende ore from the silver mines of Bohemia. Pitchblende contained uranium ore, along with a mix of other interesting materials.

On an overcast day, he placed the uranium in an enclosure with the covered photographic plate. While he did not expect to see any marking on the photographic plate, when it was developed, he found that uranium not exposed to sunlight still left a mark! He correctly concluded that the uranium emitted radiation that was not associated with X-rays.

Becquerel discovered radioactive emissions from uranium on February 26, 1896, and reported his findings soon after at the French Academy of Sciences.

Research of Marie and Pierre Curie

At the time of the discoveries of X-rays and radioactivity, Marie Curie was seeking a topic for her PhD research in Paris (Figure 1.2). She decided to study the properties of uranium using experimental techniques developed by her husband, Pierre Curie.

One of her sources of uranium was pitchblende ore. After extracting the uranium from the ore she found that the residue was still emitting considerable radioactivity.

She also systematically examined the known elements in search of other radioactive specimens and found the element thorium to be radioactive. After its elimination from the pitchblende, there were still significant radioactive emissions.

This mystery led the Curies to discover two new elements: polonium and radium that were present in the pitchblende. They shared the Nobel Prize in Physics in 1903 with Becquerel for their work on radioactivity and in 1911 Marie Curie received a second Nobel Prize, this time in Chemistry, for processing pure radium.

Frederic Soddy

A major contributor to the understanding of radioactivity was English chemist Frederick Soddy. Among his contributions was his study of the heat produced by radium. By capturing the heat in an insulated container Soddy was able to calculate the energy

Figure 1.2 *Marie Curie (1900).*
Reproduced from Library of Congress, Prints & Photographs Division, the George Grantham Bain Collection, reproduction number LC-DIG-ggbain-07682.

contained in a given amount of radium. To do this he conducted measurements of the very slow decline in the rate of energy release from radium. Soddy had a good understanding of that rate, which we know today declines by one half its energy in 1600 years.

Soddy not only contributed to the fundamental understanding of radioactivity, for which he received the Nobel Prize in Chemistry in 1921, but he was active in speaking and writing about atomic physics developments for the public.

An important contribution to public discourse was his article "The Energy of Radium" published in *Harper's Magazine* on November 30, 1909. Soddy wanted the public to know that the basic understanding of the available energy in the universe had changed due to these observations of energy emerging from radioactive elements:

> These discoveries have thus revealed a new and hitherto unsuspected store of energy in nature, resident in the structure of the elementary atoms themselves which is many hundred thousand times greater than the stores contained in the most energetic kinds of matter previously known.

He also states, ". . . there is imprisoned energy many hundred thousand times greater than is obtainable from the same weight of coal. One could carry in a pint bottle enough to drive a Mauretania (at the time, the world's largest ocean liner) around the world."

Soddy then bemoans the fact that man cannot release this energy from radium and other radioactive materials faster than its excruciatingly slow pace, such that "even the glow worm furnishes a more practically useful source of energy."

In the same article Soddy addresses the obvious question of why, given the age of the solar system, we still find radium present in significant amounts despite its constant transformation into another form.

He explains that there is a resupply of radium from the larger amounts of available uranium on earth via the process of radioactive decay. It was understood in 1909 that in the process of radioactive transformation some atoms expelled an energetic mass that was equivalent to a helium atom, the second element on the periodic table. This core of the helium atom is known as an alpha particle.

As understood by Soddy and his contemporaneous scientists, all of whom were studying radioactivity, there was a sequence of radioactive transformations, starting with uranium that undergoes transitions, first to thorium, which was then the parent of radium. The sequence continues with radium emitting an alpha particle to produce radon, which emerges in a gaseous form. The next element in the sequence is polonium produced by radon. The sequence ended when polonium emitted an alpha particle with the emergence of lead (Figure 1.3). In summary then:

Uranium → Thorium → Radium → Radon → Polonium → Lead

Each of these transitions then took place with the parent atom releasing a mass and charge equivalent to the core of the helium atom. The concept of the atom as a fundamental building block for nature that was fixed and immutable was thus revealed as an unstable and explosive object spewing vast amounts of energy into the world, albeit slowly.

This cascade of elements starting with uranium and ending with lead was suggestive of the claims of alchemists of the sixteenth through eighteenth centuries who believed

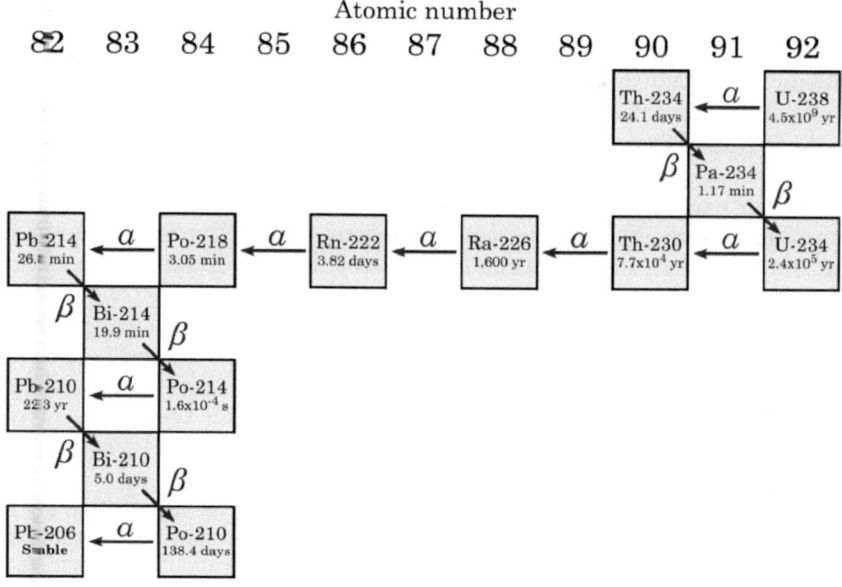

Figure 1.3 *The Uranium Decay Series. Shown are the main alpha and beta decays. A uranium-238 nucleus starting at the top right can transmute over the course of a few billion years into a stable lead-206 isotope seen in the bottom left.*

Reproduced courtesy of Alex Wellerstein (Stevens Institute of Technology, Hoboken, NJ, USA); adapted from U.S. Geological Survey (2004). Uranium. Reston, VA, USA: USGS. https://pubs.usgs.gov/of/2004/1050/PalosVerdesRn.htm#uranium.htm

that transformations of metals could be initiated that would change lead into gold. The discovery of radioactivity and the observation of uranium transforming in stages into lead challenged the conceptual foundations of science that had evolved over centuries.

These observations triggered enormous excitement in the minds of those close to the experiments. Soddy was unique in this group as not only a participating scientist but as a voice that made these astonishing discoveries known to the world with lectures, articles, and books.

H. G. Wells

At the end of the nineteenth century, while X-rays and radioactivity were being revealed in the laboratories of scientists, an English writer was using the revelations of science to establish a new genre of literature: science fiction. Wells's first novel *The Time Machine* was published in 1895 and was followed in quick succession by *The Island of Dr. Moreau* in 1896, *The Invisible Man* in 1897, and *The War of the Worlds* in 1898. When Frederick Soddy was popularizing fantastic, hitherto unsuspected natural phenomena revealed by the study of radioactivity, H. G. Wells was a well-established writer. With Soddy's exposition about radium in *Harper's Magazine* in 1909, Wells was undoubtedly aware of the possibilities for creative science fiction in these new revelations.

Wells was attracted to this scientific work and the futuristic perspective presented by Soddy and used Soddy's exposition of great energy contained in uranium and radium as the basis for themes presented in *The World Set Free* published in 1913. Early in this book, a professor named Rufus delivers a lecture to an overflow audience in Edinburgh on radium and radioactivity. In the audience there is an intense, "chuckle-headed, scrub-haired lad from the Highlands . . . drinking in every word, eyes aglow, cheeks flushed, and ears burning."

Rufus declares that ". . . radium, which seemed at first a fantastic exception, a mad inversion of all that was most established in the constitution of matter, is really at one with the rest of the elements [. . .] Radium is an element that is breaking up and flying to pieces." Rufus goes on by showing a little bottle containing about 14 ounces of Uranium oxide. He states:

> . . . the atoms in this bottle there slumber at least as much energy as we could get by burning a hundred and sixty tons of coal. If at a word, in one instant I could suddenly release that energy here and now it would blow us and everything about us to fragments; if I could turn it into the machinery that lights this city, it could keep Edinburgh brightly lit for a week. [. . .] But at present no man knows, no man has an inkling of how this little lump of stuff can be made to hasten the release of its store. [. . .] Suppose presently we find it possible to quicken that decay? [. . .] We should not only be able to use this uranium and thorium; not only should we have a source of power so potent that a man might carry in his hand the energy to light a city for a year, fight a fleet of battleships, or drive one our giant ocean liners across the Atlantic.

Later in the book, the scrub-haired lad becomes a leader in developing mechanisms to release the energy contained in uranium.

So here we have it! In *The World Set Free*, H. G. Wells clearly articulates the promise of atomic energy based on Soddy's measurements of radioactive decay. This is science fiction at its best. The novel provides a literary presentation of what might be possible in the real world based upon an understanding of scientific facts. In contrast, other books by Wells (e.g., *The War of the Worlds*) envisioned the fantasy of an alien invasion of earth. Here, the story is made plausible using science to create context. In science fiction, science is an essential element in the causation of events, while in science fantasy, science is used to create an environment in which actions take place that are impossible.

While Wells, who died in 1946, lived to learn of the devastating effects of this rapid release of internal energy at Hiroshima and Nagasaki (an outcome that he also predicted), he did not live to see his vision of atomic power with its capability of lighting a city come to fruition.

What Wells accomplished was to take the knowledge acquired by Berquelle with X-rays, together with Roentgen's, the Curies', and Soddy's understanding of radioactivity, and construct a clear picture of the potential awaiting the release of the stored energy within atoms.

The 1913 publication *The World Set Free* did not initiate an immediate quest for the key that could unlock this stored energy. However, it did provide a reasonable framework for such an undertaking. When the key was found and a process for unlocking that energy came into focus 27 years later, the world was not surprised.

James Chadwick and the 1932 Discovery of the Neutron

During the first three decades of the twentieth century much was known about the structure of atoms, but a fundamental mystery persisted. It was understood that the core of atoms contained positively charged protons that were far more massive than electrons, but these protons could not account for the mass of atoms alone. For example, the helium atom had a charge of two but the mass of the helium atom was four times that of a proton.

Similar discrepancies existed for all other elements. What accounted for this additional mass? The next theory was that a massive neutral particle existed in the nucleus. A possible model for this massive particle was suggested as a proton containing an embedded electron.

In the 1920s several scientists explored this and other possibilities by bombarding samples using the alpha particles emitted in radioactive decay. Uranium, thorium, radium, radon, and polonium were all known sources of alpha particles.

Among those who took up these investigations was avid British scientist James Chadwick, whose passion for scientific inquiries was challenged when he was unfortunately visiting Germany and imprisoned at the start of the First World War. He managed to secure radioactive toothpaste that was on the market in Germany to conduct some basic experiments while in prison.

Around 1930, while at Cavendish laboratory in England, Chadwick became interested in experiments conducted in Germany in which thin metal beryllium foils had been bombarded with polonium alpha particles. Chadwick refined these experiments to systematically determine the nature of the rays that were observed coming from the beryllium.

Key to his observations was the use of a gas-filled chamber that contained atoms that would break apart and establish a current when energetic charged particles entered the chamber (Figure 1.4). This ionization chamber was a standard instrument in physics laboratories of the day.

Chadwick found that when the rays emanating from the beryllium entered the ionization chamber they did not initiate a current. However, when a paraffin slab was placed between the beryllium and the ionization chamber a current was observed. By measuring the magnitude of the current and the time delays between emission of rays and initiation of a current, Chadwick was able to infer that the current was caused by protons knocked out of the paraffin by the neutral rays. He further deduced that the neutral rays consisted of particles with a mass approximately the same as that of protons. Chadwick had conclusively identified neutrons!

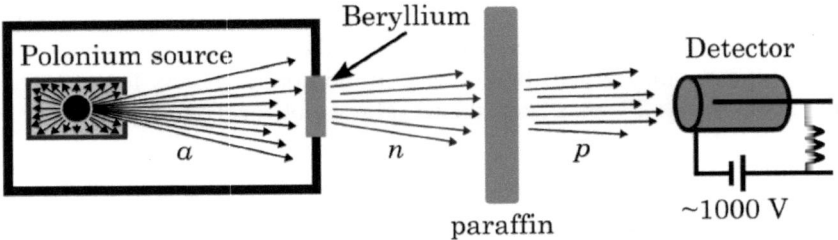

Figure 1.4 *Discovery of the neutron experiment.*
Reproduced courtesy of Alex Wellerstein (Stevens Institute of Technology, Hoboken, NJ, USA).

A Key to Quicken the Release of Energy

The discovery of neutrons meant that scientists now possessed the key to unlock the potent source of atomic power. However, few interpreted the discovery as a means to harness an energy source. Instead, neutrons were widely seen as building blocks that could be used to expand the periodic table and add new elements to it.

The periodic table organized all elements according to their charge and chemical properties. This natural order reveals important properties of the elements that had first been observed by Dmitri Mendeleev in 1869. There were 63 elements in the first periodic table and by 1940 the number had grown to 89. However, scientists were puzzled as to why uranium, with an atomic number of 92, remained the element with the largest atomic charge (Figure 1.5). With the availability of neutrons, laboratories in Rome, Paris, and Berlin strove to see if atoms more massive than uranium could be fabricated by bombarding the element with neutrons. Scientists hoped that the results of these experiments would be elements with atomic numbers 93 and possibly 94.

Taking a different perspective, Hungarian scientist Leo Szilard, who had read *A World Set Free* in 1932, immediately saw the discovery of the neutron as a way to unlock the energy source described in Wells's novel. Szilard knew Wells and was aware that the fictional account of nuclear energy was based upon Soddy's solid scientific work. Szilard became obsessed with the thought that neutrons might penetrate the charged core of atoms to trigger new, as yet unseen, reactions that could release energy stored there.

Szilard, who had studied with and gained the respect of Albert Einstein, was a brilliant but eccentric scientist who did not pursue a conventional career path. He did not seek long-term employment with an academic institution but rather moved about in an independent manner and spent long periods pursuing his own intellectual interests. One of his habits was to organize his days to include time spent hours soaking in a bathtub.

In 1933, while engaged in research work in Germany, Szilard became convinced that Hitler was a threat to world society. At a personal level, he sought safety by emigrating to England. While contemplating the possibilities of nuclear energy, he experienced an epiphany during a stroll on London streets: if a neutron could release energy from a nucleus and at the same time liberate two or more neutrons in the same reaction, then

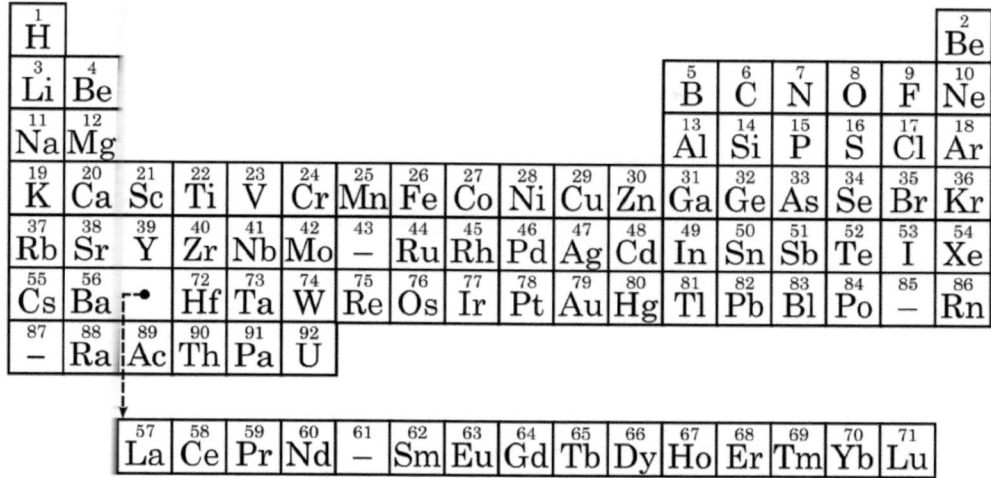

Figure 1.5 *The 1940 periodic table. Elements with dashes were yet unnamed. Along with the lack of transuranic elements, note that the positioning of thorium, protactinium, and uranium is incorrect (they were not underneath hafnium, tantalum, and tungsten, but rather cerium, praseodymium, and neodymium).*

Reproduced courtesy of Alex Wellerstein (Stevens Institute of Technology, Hoboken, NJ, USA), following Seaborg's XBL_769-1060 by Lawrence Berkeley National Laboratory.

might it be possible to achieve an explosion caused by the *two* neutrons promoting the liberation of *four*, with successive doublings (Figure 1.6)? This neutron cascade (i.e., chain reaction) was a mechanism he foresaw as a process to release the energy stored in atoms (Figure 1.7).

Using the concept of the chain-reaction process, Szilard next sought opportunities for experimental verification. Following the work of Ernest Rutherford, who had bombarded beryllium with alpha particles to yield carbon plus a neutron, Szilard saw beryllium as a rich source of neutrons. He considered that, by bombarding beryllium with neutrons, it might be possible to release energy in a reaction that emitted two neutrons.

He sought opportunities to explore this and other related nuclear transformations by approaching university-based research groups and corporate laboratories in England. In his conversations with other scientists, he was severely handicapped by his great fear that German researchers might successfully pursue this path and make nuclear weapons available to Hitler. While eager to engage collaborators with whom he could explore these new possibilities, he remained guarded to avoid revealing information that might reach German scientists.

Szilard's creativity brought forth many original ideas. Generally, scientists viewed patent applications as contrary to the idea of open exploration, that is, the essence of pure science. Unlike most scientists, he was experienced at applying for patents (an activity more frequently pursued by engineers), including having applied jointly in the

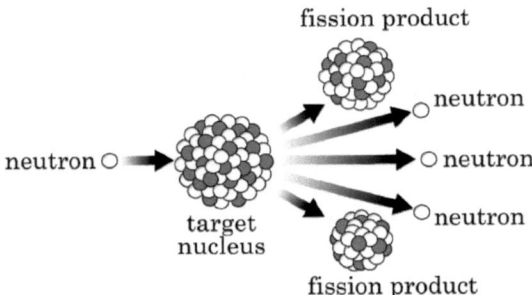

Figure 1.6 *Nuclear fission reaction. The incident neutron at left is absorbed by the target heavy nucleus. This in turn splits (fissions) into two fission products of roughly equal mass as well a several neutrons.*

Reproduced courtesy of Alex Wellerstein (Stevens Institute of Technology, Hoboken, NJ, USA). Adapted from MikeRun (2017). Wikimedia Commons. https://commons.wikimedia.org/wiki/File:Nuclear_fission_reaction.svg, under a Creative Commons Attribution-Share Alike 4.0 International (CC BY-SA 4.0).

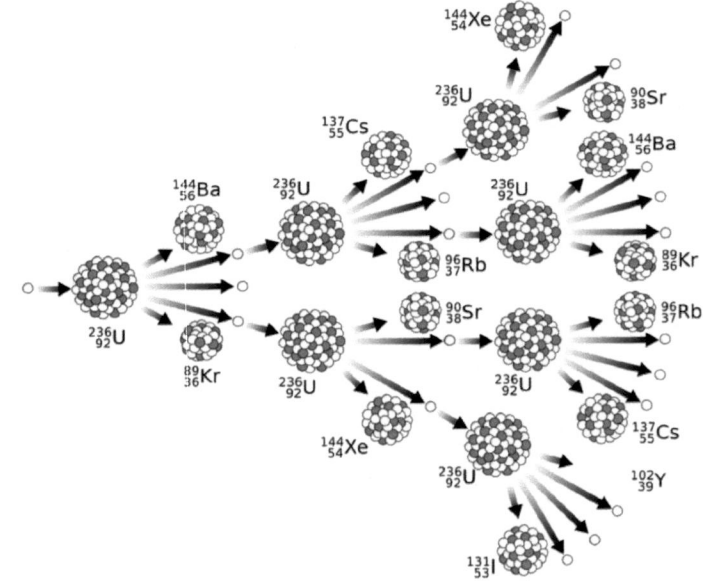

Figure 1.7 *Nuclear chain reaction in which one fission reaction begets a cascade of multiple follow-on fission reactions.*

Reproduced courtesy of Alex Wellerstein (Stevens Institute of Technology, Hoboken, NJ, USA). Adapted from MikeRun (2017). Wikimedia Commons. https://commons.wikimedia.org/wiki/File:Nuclear_fission_reaction.svg. under a Creative Commons Attribution-Share Alike 4.0 International (CC BY-SA 4.0 .

United States with Albert Einstein in 1930 for an innovative refrigerator design. This was the only patent attributed to Einstein who had worked in a patent office prior to his success as a physicist. Other Szilard patents included plans for: microfilm, a cyclotron, an electron microscope, and an accelerator to speed up charged particles. Clearly it was true to form for Szilard to apply for a patent for releasing atomic energy via chain reaction, and he filed a fifteen-page application on March 12, 1934. The application, revised several times, was speculative, since he could not describe a clear path leading to energy release. Of the several possibilities outlined, that involving Beryllium was wrong because it relied on incorrect published data about Beryllium. Szilard's patent, which has been criticized as garbled and confused, reflects both his incomplete knowledge of what might work as well as his reluctance to provide clear ideas that might help Nazi researchers.

After filing his application, Szilard engaged several scientists in efforts to establish research programs to implement a chain reaction with an associated large-scale release of energy. The most significant of these was a meeting Rutherford on June 4, 1934, three months after the patent application. Szilard was just as conflicted during that meeting as he was preparing the patent. While he told Rutherford of the patent, he only spoke about alpha particle experiments—not about the use of neutrons—and his request to Rutherford that he be hired to conduct breakthrough experiments was futile. Rutherford quickly ended the conversation.

Szilard went on to broach his ideas to Hugo Hirst, the director of the British General Electric Company research programs in England, but he again limited the conversation, this time to a discussion of medical isotope production. While he succeeded in obtaining an offer from the future Nobel Laureate George Paget Thompson to lead a research program at the University of London, he turned it down. Thompson expected him to publish his results and to not pursue a secret program.

During the summer of 1934 Szilard engaged in medical isotopes research at St. Bartholomew's Hospital with Thomas A. Chalmers. They discovered an important process in which radioactive elements could be isolated following bombardment of a compound in which ionization was triggered.

For the next year, dead-end meetings at Oxford, New York University, and Manchester University led to his offering the patent to the War Office on September 16, 1935. Since the Army would not agree to pursuing the patent in secret, he turned to the British Admiralty, which issued a certificate of secrecy on March 26, 1936.

Szilard overcame neither the resistance of scientists to pursue research in secret nor the skepticism of many (including Enrico Fermi) that vast amounts of energy could be released by neutron bombardment.

Fermi's Unseen Discovery of Fission

Remarkably, Enrico Fermi, one of the greatest scientists of the twentieth century, who led a comprehensive search starting in 1934 using neutrons to bombard existing elements, incorrectly concluded that he had discovered elements 93 and 94. He named them ausenium (after Ausonia, the Greek name for Italy) and hesperium (after Hesperia, a poetic name for Italy), respectively. In reality, he had triggered the breakup of uranium with the release of stored energy but did not realize it. The announcement of what was thought to be discovery of new elements was hailed by the press. Science historian Ruth Lewin Sime reported that an Italian newspaper claimed the discovery as proof of Italy's restored grandeur under the Fascists. She also referenced another false report that Fermi had presented a small vial of element 93 to the queen of Italy.

The mistakes occurred due to a prior mindset that discovering these elements was a given, and that the actual experiments resulting in the breakup of uranium yielded radiation that appeared to originate in heavier elements. Fermi mistook radiation emitted by lighter elements as the radiation he anticipated from elements 93 and 94. While he systematically explored the results of exposing neutrons to more than sixty elements, Fermi had made many important discoveries that led to his receipt of the Nobel Prize. He learned of his mistake with ausenium and hesperium too late to eliminate references to those discoveries in his Nobel acceptance speech, but his many valid discoveries nevertheless justified his award.

Sime noted that "No one suspected that a mild-mannered neutron could sidle up to a uranium nucleus and trigger an explosive disintegration." While Soddy and others conclusively demonstrated that explosive potential energy existed in the uranium nucleus, the neutron was overlooked as being the key to unlocking that store.

Hahn Observes (and Meitner Explains) Fission

Fermi witnessed but didn't recognize fission, and working with Paul Savitch in Paris, Irene Curie (daughter of Marie and Pierre) conducted experiments similar to Fermi's but gained no new insights. However, the breakthrough came at the Berlin Kaiser Wilhelm Institute. The team working there was led by nuclear chemist Otto Hahn and nuclear physicist Lise Meitner (Figure 1.8).

Meitner was exceptional. The second woman to receive a doctorate from the University of Vienna, she was the first woman to be elevated to the position of full professor of physics in Germany. While she was born in 1878 in Austria to Jewish parents, she converted to Lutheranism and was baptized in 1908. She considered herself a Christian and was not subjected to the increasingly vicious oppression of Jews in 1930s Germany. While many Jewish scientists left Germany at that time, most notably Einstein (who emigrated in 1933), Meitner remained in Berlin.

However, she was no longer safe when Hitler annexed Austria in March of 1938 and Germany declared that anyone whose grandmother was Jewish would be classified as Jewish. Finding herself in great danger, she was able to escape from Germany to Sweden via the Netherlands. Her emigration occurred at a critical time in the experimental program that she and Otto Hahn had undertaken, in which they were examining the results of bombarding uranium with neutrons. Hahn was an expert in conducting these experiments and in organizing the results. In this he was assisted by Fritz Strassman.

Hahn's careful analysis of the data found evidence of light elements (e.g., barium) emerging from the uranium target after neutron bombardment, which was unexpected

Figure 1.8 *Otto Hahn and Lise Meitner in Berlin (1912).*
Unknown author. Reproduced from Churchill Archives Centre, Churchill College, Cambridge. Reference: GBR/0014/MTNR 8/4/1.

and puzzling. Hahn was engaged in daily correspondence with Meitner, who reviewed the experimental results from Sweden, where she had been joined by her nephew Otto Robert Frisch (who was also a theoretical physicist). She had an insight into the dynamics of the neutron–uranium interaction that led to development of a model for the breakup of uranium into two smaller fragments. She and Frisch called this process "fission."

After communicating this insight to Hahn, he published his results using the Meitner analysis but not mentioning her part in the research. Instead, he submitted the research results with Fritz Strassman as co-author on December 22, 1938 and omitted Meitner's name; he did not want to associate himself with a Jewish collaborator. Meitner and Frisch then published a separate theoretical interpretation of the results in *Nature* in March of 1939. Sadly, Hahn's cowardice led to Meitner's not being included in the Nobel Prize awarded to Hahn in 1944.

While Albert Einstein referred to Meitner as the German Marie Curie, and it is an appropriate comparison regarding the significance of the scientific accomplishments of the two women, the parallels fail when examining the recognition each received. Marie Curie received two Nobel Prizes, and Lise Meitner none. Meitner could be viewed in history as the most under-recognized woman scientist in history.

News of Hahn's publication spread rapidly among scientists around the world. Robert Oppenheimer, who was then a professor at the University of California, Berkeley, immediately laid out the parameters for a bomb design. Szilard, when he confirmed Hahn's findings at New York's Columbia University, said that the world was headed for sorrow.

Fission was discovered just before the start of the Second World War II, with the result that all efforts around this new knowledge were around producing a bomb. The promise of nuclear energy to advance society was placed on hold.

Further Reading

Antoine H. Becquerel, *On Radioactivity, New Property of Matter*, Nobel Lecture, December 11, 1903.
Becquerel's account of his discovery of radioactivity.

Frederick Soddy, "The Energy of Radium," *Harper's Magazine*, December 1909, 52–59.
Soddy's calculation of the energy available if radium were to release its energy all at once instead of over an extended time period. This calculation anticipated the potential of nuclear energy and nuclear weapons.

H. G. Wells, *The World Set Free*, Macmillan & Co., 1914
Wells uses the insights of Frederick Soddy to envision a world that experiences nuclear war and uses nuclear energy.

2

US Developments: 1940–1945

*It is such a supreme folly to believe that nuclear weapons are deadly only if they're
used. The fact that they exist at all, their presence in our lives, will wreak more havoc
than we can begin to fathom. Nuclear weapons pervade our thinking. Control our
behavior. Administer our societies. Inform our dreams. They bury themselves like
meat hooks deep in the base of our brains. They are purveyors of madness.*

— Arundhati Roy[1]

Emergence of Nuclear Technology

Nuclear energy was born from America's Manhattan Project, a highly secretive program
that sought to unleash the newly understood physics of nuclear fission for war, rather
than harness that knowledge for peace. The dreams of the scientists working on the
project to explore potential peaceful uses would have to wait, as such explorations took
a back seat to the war effort in the 1940s. Spanning the years of the Second World War,
the project culminated in the first and only wartime use of nuclear weapons to date, when
the United States dropped the new class of bomb on the Japanese cities of Hiroshima
and Nagasaki in early August 1945.

Driving hard for a formal weapons program was Hungarian-born physicist Leo Szilard, who foresaw the possibility of fission weapons and feared that Nazi Germany
would be the first to acquire them. Aware that as an unknown immigrant his call for
action would receive little attention, he recruited his friend Albert Einstein, already a
world-famous physicist, to approach President Franklin Roosevelt with an appeal for
the United States to pursue the development of a fission bomb. Szilard drafted a letter
for Einstein to send to Roosevelt. Einstein, who was a pacifist, became convinced that
the Nazi danger was so great that he was willing to revise and send the letter to Roosevelt.
The letter dated August 2, 1939, stimulated action at the White House. Though at first
this action was quite limited, it intensified as war loomed.

The Manhattan Project evolved rapidly during the early years of the war, beginning
with an acceleration of research at American and British universities, followed by establishing the Office of Scientific Research and Development, headed up by Vannevar Bush,
and the guiding S-1 committee, composed of Harold C. Urey, Ernest O. Lawrence,

---

[1] Arundhati Roy, "The End of Imagination", *My Seditious Heart: Collected Nonfiction*. Haymarket Books,
2019, p. ●.

Nuclear Energy. Edward A. Friedman, Oxford University Press. © Edward A. Friedman (2025). DOI: 10.1093/9780198925811.003.0002

James B. Conant, Lyman J. Briggs, E. V. Murphree, and Arthur Compton, leading American physicists, chemists, and engineers. The effort was formally dubbed the Manhattan Project in 1942, when the US Army Corps of Engineers in the non-descript "Manhattan Engineer District" consolidated research efforts and began the top-secret push for the large-scale design and production of the components needed for uranium and plutonium bombs. The military aspects of the program were organized under the hard-driving leadership of General Leslie R. Groves, a West Point graduate with experience in military engineering and construction, including that of the Pentagon. Overseeing the construction of the Pentagon, which was completed in 1943, was a major achievement since it had the largest floor area of any office building in the world. That was a distinction that it maintained for 80 years, until the Surat Diamond Bourse surpassed it in Gujarat, India in 2023.

Broad scientific control on the design and construction of the weapons was given to Dr. J. Robert Oppenheimer, who occupied a leadership position as head of a theoretical physics group at the University of California, Berkeley. Oppenheimer had obtained his doctorate in physics from the University of Gottingen in Germany in 1927 and was one of the few American physicists who interacted with Gottingen—at the time the world's leading center for the study of atomic and nuclear physics. Oppenheimer was the best choice to lead the weapons development program given his expertise in the field and the fact that he was a rare native-born US citizen in a US physics community that was increasingly dominated by Europeans.

Although the Manhattan Project sought to build nuclear weapons, the unity of top research institutes and military–industrial capability set the stage for the successful development and deployment of nuclear power reactors after the war.

One of the earliest breakthroughs of the Manhattan Project preceded America's direct involvement in the war. On December 14, 1940, while working at the University of Chicago, Glenn T. Seaborg discovered a new element, which he named plutonium. A few months later he determined that this new element could sustain a fission chain reaction resulting in a greater energy output than that of uranium. It was this discovery that set the stage for the Manhattan Project's dual focus on uranium and plutonium.

When Japanese forces attacked Pearl Harbor on December 7, 1941, the organizational infrastructure was in place to support Italian American physicist Enrico Fermi and a team of leading scientists in establishing the Manhattan Project's Metallurgical Laboratory, or MetLab, at the University of Chicago in early 1942. Before embarking on a program to build a bomb there needed to be experimental proof that a uranium fission chain reaction could be sustained.

Chicago Pile-1

To attain this proof, Fermi and the MetLab scientists occupied an unconventional location: a racket court under the viewing stands of the university's original Stagg Field. There they built the world's first artificial nuclear reactor, the Chicago Pile-1. Civilian and military leaders had strong misgivings about the possibility of a runaway reaction taking place in this urban setting. However, the site provided the space, infrastructure, and access to world-class scientists needed to carry out the task while maintaining secrecy.

To achieve a chain reaction using natural uranium, 330 tons of ultra-pure graphite were procured, which itself required the development of novel purification techniques. The graphite served the purpose of slowing the neutrons to a velocity that optimized triggering a chain reaction, minimizing the capture of neutrons within the graphite's remaining impurities. Fermi called the reactor "a crude pile of black [graphite] bricks and wooden timbers" in which 50 tons of uranium metal and uranium oxide were strategically placed. Since it was designed to operate at exceedingly low power levels, there was no effort to incorporate a cooling system or radiation shielding seen on future reactors (Figure 2 1).

On December 2, 1942, just short of one year after the Pearl Harbor attack and America's involvement in the Second World War, the MetLab scientists gathered in the racket court to witness the test of the completed reactor. After a cautious withdrawal of the cadmium control rods, the Chicago Pile experienced a self-sustaining chain reaction. This condition was maintained for four and one-half minutes at a power level of one-half a watt before it was shut down with the reinsertion of the control rods. It was a success. The physicists assembled for this transition into the age of atomic energy opened a bottle of Chianti and sipped toasts from paper cups. Compton, who managed the Chicago

Figure 2 1 *An illustration of the first critical nuclear reaction. Photograph of an original painting by Gary Sheehan (1957).*

Reproduced from Records of the Atomic Energy Commission. National Archives Identifier: 542144. https://catalog.archives.gov/id/542144.

research program, reported this success to Conant, whose office in Washington, DC oversaw the Manhattan Project, with a telephone call that used an impromptu code:

> Compton: The Italian navigator has landed in the New World.
> Conant: How were the natives?
> Compton: Very friendly.

Manhattan Project

To take advantage of these insights from atomic physics, the Manhattan Project accelerated into a massive undertaking that exploited the industrial capacity of the United States as well as the research capabilities of the United Kingdom and Canada. What was now needed was the production of enough fissionable uranium and plutonium for bomb production.

One of the three main Manhattan Project industrial sites was placed along the winding Clinch River at Oak Ridge, Tennessee, which provided an ideal mix of seclusion, accessibility for large trucks bringing materials to the site, and availability of hydropower. Known as the Clinton Engineer Works, the site employed tens of thousands of people, requiring the development, from nearly uninhabited wilderness, of the nearby Oak Ridge town. By 1944, the town was home to around 75,000 people and entirely under management by government contractors. These contractors provided doctors, dentists, recreation, schools, cafeterias, and churches to residents, albeit quite unequally in a segregated environment.

The site housed an early reactor to create plutonium, but its primary mandate was to solve the problem of uranium enrichment by increasing the relative quantity of fissionable isotopes in natural uranium. The isotope of uranium that undergoes fission is U-235, but this isotope makes up just 0.7% of naturally occurring uranium, with the rest being almost entirely non-fissile U-238. Thus, one of the primary challenges of the Manhattan Project was to solve, on a large scale, the problem of separating the uranium isotopes to achieve high enough concentrations of fissile U-235 to use in bombs. The concentration needed was over 90%. Huge facilities were built at Oak Ridge for this purpose, with their various processes taking advantage of the slight weight difference between U-235 and U-238. Both isotopes have nuclei containing 92 protons, but they differ in the number of neutrons, with U-235 having 143 neutrons and U-238 having 146. While this is less than 1% difference in mass, it was sufficient to provide the basis for separation of the isotopes.

The separation of uranium isotopes was an engineering problem without precedent. An early method of isotope separation considered was the use of centrifuges. For this, the uranium would first need to be combined with fluorine to create uranium hexafluoride, a gaseous form of uranium. This gas, when placed in a spinning centrifuge was expected to allow the two uranium isotopes to separate, since the slightly heavier form moved in a different trajectory than the lighter form. High-speed centrifuges had been developed at a lab scale in the late 1930s. However, scaling up the centrifuges and using them for uranium separation created manufacturing challenges and lower-than-expected yields of U-235. As early centrifuge experiments were running, another potential method called gaseous diffusion was gaining favor among scientists In December of 1942, a final approval committee chaired by MIT chemical engineer Warren K. Lewis and membered by three representatives of the DuPont Chemical Company, the primary contractor for the Manhattan Project, recommended that gaseous diffusion be pursued at Oak Ridge owing to a higher perceived chance of success.

Gaseous diffusion used membrane barriers that allowed the lighter isotope in the gaseous uranium hexafluoride to pass through its microscopic openings more readily than the heavier isotope. Based on preliminary research from Columbia University, additional research and development on the gaseous diffusion facility continued long after construction began in 1943, with the production of a suitable barrier being a dominant and persistent challenge that was only resolved in early 1945.

To obtain usable amounts of U-235, the individual diffusion process needed to be repeated nearly 3,000 times. An enormous building was constructed to house this operation, which at the time was the largest building in the world by footprint, with a length of 875 yards (800 m) and a width of 1,000 yards (914 m). Ten thousand workers were employed at this top-secret facility, which at its peak consumed a tenth of all energy used in the United States. Oak Ridge was able to secure this vast amount of electricity from the neighboring Tennessee Valley Authority.

A secondary method of isotope separation explored at Oak Ridge was based on research by Ernest Lawrence, a physicist at the University of California, Berkeley. This form of separation subjected ionized uranium hexafluoride to electronic forces that resulted in differing trajectories of the isotopes. For this purpose, the Y-12 electromagnetic separation plant was designed and constructed and finally, with great effort, entered operation in 1945.

Obtaining the vast amounts of conductor material required for the electromagnets became a significant challenge with a unique solution. Shortage of copper amid the broader war effort meant that the project ultimately negotiated to borrow nearly 15,000 tons of silver from the US Treasury in the form of 1000-oz silver bars, which were melted down to form the large magnetic coil "racetracks" at Y-12, with meticulous accounting enforced by Groves.

Unsure of which method of isotope separation would ultimately prove successful, Manhattan Project leaders pursued both gaseous diffusion and electromagnetic separation with nearly equal intensity, making use of thousands of businesses and contractors within the private industrial sector under a tight secrecy regime. The scale of the K-25[2] and Y-12 facilities was enormous, with their construction and operation costs exceeding one billion dollars, around half of the cost of the entire project. Ultimately, the K-25 and

---

[2] The K-25 gaseous diffusion plant separated lighter hranium 235 from heavier uranium 238. The K-25 relied on over 3,000 gaseous diffusion steps utilising uranium hexafluoride gas.

Y-12 plants were used in series, with K-25 providing initial enrichment that was then used as a feedstock for the Y-12 plant. After the war, with the K-25 plant proving to be more efficient than Y-12, the Y-12 plant was retired.

Pursuing multiple bomb designs, the Manhattan Project was interested not just in the enrichment of uranium but also in the production of plutonium. The first plutonium production facility was built at Oak Ridge using the world's second nuclear reactor, known as the X-10. This reactor was a center where techniques could be developed to irradiate uranium fuel containers and then process them chemically to isolate plutonium. It used natural uranium that was air-cooled and used graphite as a moderator. The reactor went critical on November 4, 1943, and produced the first useful amounts of plutonium in early 1944. Unlike Chicago Pile-1, the X-10 was the first reactor to achieve continuous operation. Its power level, which was enhanced during its first year of operation, reached 4,000 kilowatts.

Expanded plutonium production was pursued at a site in Hanford, Washington. Similar in size to Oak Ridge, the site employed 51,000 people. Under control by DuPont, the nearby town of Richmond boomed from around 250 residents to 15,000 within one year of General Groves's approval of the site in January of 1943. Segregated and prohibited from living in the "idyllic" Richmond, 15,000 Black workers were housed in squalid barracks outside of town. The project and its workforce constituted the fourth-largest town in Washington.

At Hanford, nuclear reactors were built to provide a massive exposure of natural uranium to neutrons, thus generating plutonium (Pu)-239 from U-238 nuclei that had absorbed a neutron. The fuel rods in these reactors therefore contained a mixture of plutonium and isotopes of uranium that were processed chemically to separate the plutonium for use in weapons.

Three such graphite reactors, informed by experience with X-10, were constructed at Hanford, which had dimensions of 36 × 36-foot cross section and a depth of 28 feet. The water-cooled uranium fuel was distributed in 60,000 containers. Four chemical processing plants were built to separate the plutonium. The first of these reactors became operational on September 26, 1944.

While Oak Ridge became the center for enrichment of U-235 that fueled the Hiroshima Bomb and Hanford became the center of production of plutonium that fueled the Nagasaki bomb, the heart of the Manhattan Project was the Oak Ridge Laboratory, where the basic science was studied for the design and fabrication of those bombs. It was Oppenheimer, whose family owned a ranch in New Mexico, who urged that a center be established where scientists could pursue development of the bombs in an isolated environment. He knew from his personal experience vacationing in the region that the natural beauty of the region could stimulate reflection on the awesome challenges that they were contemplating. He also prevailed upon General Groves the necessity of allowing working groups that encompassed the entire team of scientists without restricting discussions to compartmentalized units. He wished to engage all the scientific minds at the laboratory in the problem solving that lay before them.

The personnel living and working at Los Alamos, known by its code name Project Y, totaled over 10,000 by the war's end. Of these, no more than several hundred were scientists. The population included technicians, military security personnel, families, support staff, and those needed to care for the health and wellbeing of the civilian population, including those running schools for children and managing clinics.

Histories of the first atomic bombs (i.e., the uranium bomb dropped on Hiroshima and the plutonium bomb dropped on Nagasaki) often focus on their construction by the scientists who worked at Los Alamos. However, it was not only scientists, but also the significant industrial capacity of US corporations that enabled the Manhattan Project to succeed and help end the war in the Pacific in 1945.

At its peak in June 1944, the Manhattan Project employed about 129,000 workers, of whom 4,500 were involved in construction, 40,500 were plant operators, and 1,800 were military personnel. Due to high turnover, over 500,000 people worked on the project at some point. The scale of the endeavor is virtually unmatched in US history, but its speed was equally impressive. Only 32 months elapsed between the time a chain reaction was observed at the Chicago Pile-1 in December 1942 and when bombs were detonated in Japan in August 1945.

Trinity

When the team at Los Alamos was confident that they could construct U-235 and Pu-239 bombs, it was decided that at least one bomb test was required before using these devices. The test required was for the plutonium bomb as its design was more complex than that of the uranium bomb. Characteristics of U-235 supported development of a "gun" type bomb in which half of the uranium could be fabricated into the bullet of a modified artillery gun and half fabricated into the target located in the artillery gun assembly. This construction is so simple that it was judged that nothing could go wrong with its implementation.

However, the volatility of Pu-239 is such that it would cause a fizzle reaction if used in a gun type design. Instead, the plutonium bomb design uses multiple chunks of Pu-239 distributed around the surface of a sphere.

These multiple chunks of plutonium are propelled by chemical explosives into the center of the sphere where they collide and trigger the nuclear explosion.

These multiple chemical explosions must be set off with exquisite timing so that individual chunks arrive at the same time at the center of the sphere. This high-precision timing could fail and therefore the design required a test.

The test was conducted at 5:30 a.m. on July 16, 1945, in the bombing range near the Alamogordo Army Airfield in the New Mexico desert near Albuquerque. The bomb was ignited while suspended 100 feet in the air on top of a steel tower. The yield was about 20 kilotons of TNT, close to the detonation of the plutonium bomb used at Nagasaki.

Prior to its detonation, some physicists speculated that the explosion could have resulted in a chain reaction that ignited the earth's atmosphere. While the probability of such an event was considered low, there was relief that such a catastrophe did not occur.

An unusual consequence of the Trinity explosion resulted from the intense heat to which the desert sands were exposed. The author of this book visited the Trinity site in 2020 and witnessed the vast area of desert that had been vitrified by that heat. An expanse of small nuggets of glassified sand is the only evidence that the world's first atomic explosion took place there in 1945.

Obstructing the German Bomb

Throughout the war, Allied forces were highly anxious that Germany would be successful in developing an atomic bomb. Between 1943 and 1945, the Alsos Mission, a military collaboration between the United States and Britain, sought to uncover potential developments in the German bomb project. In early 1945, Alsos operatives were successful in entering Nazi Germany and extracting leading scientists who had been working on the country's bomb project. They were taken to England and confined to a house where their conversations were recorded. Upon learning of the detonation of the nuclear weapons in August 1945, they were observed to express astonishment that the United States had achieved the fabrication of nuclear devices in the short time that had elapsed from the beginning of the war.

The German bomb project, for its part, had been seriously damaged by the destruction of a Norwegian heavy water plant in early 1943 in a series of bombing missions by Norwegian and British forces. Heavy water refers to water in which hydrogen is replaced with atoms whose nucleus contains one proton and one neutron. This isotope, known as deuterium, can be used to cool reactors without absorbing neutrons that are engaged in sustaining chain reactions. Thus, heavy water can be used as a moderator in reactors that use natural uranium fuel that contains only a small portion of fissile U-235. Germany sought to use this type of reactor for its weapons research, with heavy water planned to be sourced from a production facility in Vemork, Norway near a hydroelectric power plant, where the large amounts of energy required for this process were available.

After the destruction of the facility, the Germans attempted to move the surviving heavy water supplies to Germany via a ferry. Upon learning the ferry's route, three Norwegian resistance fighters managed to board the vessel with 8.5 kg of plastic explosives on timed fuses and sank the ferry in Lake Tinn. Partially owing to the success of these sabotage missions, Germany was never successful in developing nuclear weapons.

Entering the Public View

As the war ended, scientists took the lead in advocating civilian control of atomic energy. An intense political struggle ensued in which Szilard again played a key role. Not surprisingly, Groves was passionately engaged in efforts to maintain military supremacy over America's nuclear future. This struggle was a continuation of the conflict between these two individuals over nuclear policy matters from the very beginning of the Manhattan

Project. Groves was so incensed by Szilard's anti-authoritarian behavior that, early in the Manhattan Project, the general advocated for imprisoning Szilard as a danger to US interests. However, this post-war debate was conducted in the US Congress, where there was strong support for civilian control.

On August 1, 1946, President Harry Truman signed the Atomic Energy Act that had been sponsored by Connecticut senator Brien McMahon as chair of a special senate committee on atomic energy. In addition to creating the Atomic Energy Commission, administered by a five-member board of civilians, the Atomic Energy Act also established the National Laboratories. With their origins in the Manhattan Project, the Argonne National Laboratory was based in Illinois, the Oak Ridge National Laboratory in Tennessee, and the Los Alamos Laboratory in New Mexico. Additionally, the Brookhaven National Laboratory was established in Long Island, New York. These facilities pursued both weapons projects as well as civilian applications.

In 1945, despite much being said about the opportunities for civilian use of nuclear energy, there was little interest from government agencies, Congress, or the power sector for the development of nuclear power reactors. The government and Congress were more concerned with maintaining a global lead in nuclear weapons development, while the producers of electrical power were satisfied with large supplies of inexpensive coal. Without any commercialized technologies yet available, power companies viewed nuclear power as a far-off promise that was both expensive and inaccessible. Moreover, the details of nuclear energy generation were secret, and the power companies did not have personnel trained in nuclear engineering.

Secrecy and Spies

Leading up to the detonation of nuclear weapons on Japanese cities in August 1945, the work of the Manhattan Project was remarkably secretive. Despite thousands of workers, contractors, and scientists involved, Groves's extreme compartmentalization and authoritarian control meant that very few individuals glimpsed more than a sliver of the work being done. In 1945, *Life* magazine wrote that the more than 100,000 individuals involved in the project, "worked like moles in the dark." Penalties of more than ten years imprisonment and heavy fines were threatened for anyone disclosing the project's secrets. News media were discouraged from reporting on topics related to the activities of the project.

While the United States engaged in joint atomic weapons development with the United Kingdom and Canada, it did not share these initiatives with France or the Soviet Union. A counter-intelligence corps was established to monitor the project's security. There was special concern regarding the intentions of the Soviet Union which, while an ally, did not share the democratic values of the United States or Western Europe and was not trusted.

However, during the 1930s, communist society as it was being developed in the Soviet Union attracted the support of many intellectuals in Europe and the United States.

There were a few with that orientation who voluntarily passed on information about the Manhattan Project to the Soviet Union, most notably Klaus Fuchs.

Fuchs was the son of a Lutheran pastor living in Leipzig, Germany, who became active with the communist party in the early 1930s. He fled from Nazi-dominated Germany to the United Kingdom, where he studied atomic physics under leaders in the field at British universities. In 1941, he joined the early British atomic bomb project and began passing information about that research to Soviet military intelligence agents in the United Kingdom.

The team he worked with joined the Manhattan Project in 1943, where he became a leading theoretical physicist in the group developing the plutonium implosion-type nuclear weapon design. While working at Los Alamos, he successfully passed information about the bomb program to an intermediary who he would regularly meet in Santa Fe, New Mexico. This detailed material was delivered to the Soviet Consulate in New York, from where it was transmitted to Russia.

Just before the use of nuclear weapons in August 1945, President Truman met with Joseph Stalin in Potsdam, Germany. After announcing that the United States was developing a powerful weapon, Truman was shocked that Stalin showed little interest or surprise. Having only learned about the Manhattan Project after becoming US president just months earlier in April of 1945, Truman was unaware that Stalin was more familiar with the Los Alamos work than he was.

After the war, Fuchs returned to the United Kingdom, where he became head of the Theoretical Physics Division of the British atomic energy program. It was not until 1950 that his activities as a spy were revealed during an FBI review of wartime coded transmissions from the Soviet consular office in New York to Russia. When American authorities informed the British of Fuchs's activities on behalf of the Soviet Union, another Soviet agent named Kim Philby was serving as a UK–US intelligence liaison. Philby was able to prevent the US agents from interviewing Fuchs and was instrumental in limiting the British investigation of Fuchs's activities.

A concern for the ability of the British team to maintain secrecy led to the 1946 decision to end American collaboration on nuclear weapons development with the United Kingdom, which decided to embark upon its own nuclear weapons program in 1947.

After serving nine years in a UK prison, Fuchs returned to Germany where he served as deputy director of the East German Institute for Nuclear Physics.

Thus, we see that the secrecy with which the United States treated nuclear energy did not prevent or significantly delay the USSR from developing nuclear weapons. However, the secrecy, which limited access to nuclear information, did obstruct the development in the United States of nuclear energy technology. Both universities and corporations lacked the knowledge needed for the development of power reactors. Eventually, the transition from military applications of atomic energy to peaceful pursuits found a foothold in the US Navy's development of nuclear submarines. This pathway to commercial nuclear power would shape the course of nuclear energy around the world.

Further Reading

F. G. Gosling, *The Manhattan Project: Making the Atomic Bomb*, United States Department of Energy, 2010.
A succinct account of the Manhattan Project that is well illustrated.
Richard Rhodes, *The Making of the Atomic Bomb*, 25th Anniversary Edition, Simon & Schuster, 1995.
This definitive history of the making of the atomic bomb is an award-winning classic.

3

US Developments 1945–1960

To the making of these fateful decisions, the United States pledges before you, and therefore before the world, its determination to help solve the fearful atomic dilemma—to devote its entire heart and mind to finding the way by which the miraculous inventiveness of man shall not be dedicated to his death but consecrated to his life.

—President Dwight D. Eisenhower[1]

Military Development of Reactors for Energy

When the Second World War ended, the US military shifted their attention from bombs to transport. The US Navy had a strong interest in the development of a nuclear submarine that could stay underwater for extended periods of time. The US Air Force sought the advantages of a nuclear-powered plane that could travel great distances without refueling. In 1945, the Air Force was still an Army military unit and did not achieve independent status until 1947.

Development of a submarine reactor became an objective of the Argonne Laboratory and nuclear plane development was undertaken at Oak Ridge. These projects came under the administrative leadership of two extremely talented individuals with quite different personalities: Hyman G. Rickover and Alvin M. Weinberg. Rickover was an immigrant from Poland who graduated from the US Naval Academy, while Weinberg was a nuclear physicist who served as an administrator at Oak Ridge during the Manhattan Project. Rickover, after serving on naval vessels in the 1920s, earned a master's degree in electrical engineering from Columbia University in 1930 and, from 1939, headed the electrical section of the Bureau of Engineering. Rickover had a hard driving engineer–administrator manner, while Weinberg was more of a reflective scientist. The influence exerted by these two men determined for many decades how nuclear energy developed in the United States and much of the rest of the world.

Rickover became involved in nuclear engineering in 1946 at Oak Ridge (then the Clinton Laboratory) and in 1947 Nimitz appointed him chief of the newly created

---

[1] Dwight D. Eisenhower, "Atoms for Peace." Speech to the United Nations General Assembly, December 8, 1953.

Nuclear Energy. Edward A. Friedman, Oxford University Press. © Edward A. Friedman (2025). DOI: 10.1093/9780198925811.003.0003

Nuclear Power Division of the Bureau of Ships. It was in this position that he met Weinberg. He also enlisted the Westinghouse Electric Company to establish the Bettis Atomic Power Laboratory within Argonne that became a research and development site for naval nuclear programs.

Rickover and Weinberg took different design paths in pursuit of power reactors in meeting the challenges of their missions. Rickover's nuclear knowledge was limited to systems in which water served as the moderator, coolant, and fluid that transferred heat to electrical generators. Rickover focused the work at Argonne on pressurized-water and boiling-water reactors (Figures 3.1 and 3.2). While Weinberg was familiar with the water-based designs developed at Oak Ridge, he chose a design using a molten salt fueled and cooled reactor to power aircraft.

Both programs were successful in developing operating reactors, although the molten salt reactors had corrosion problems that impeded possible implementation. However, other issues doomed the project and eventually led to its cancellation. These included concern about potential radiation dangers if a plane were to crash; the development of techniques for aerial refueling; and emerging missile technology as an alternative strike

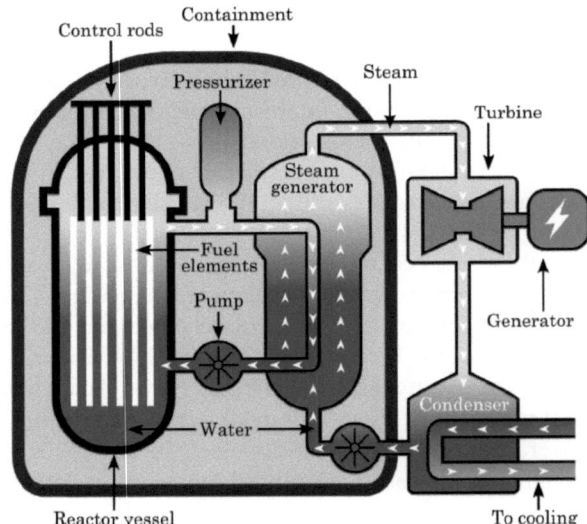

Figure 3.1 *A schematic representation (not to scale) of the basic principles of a pressurized water reactor (PWR). Once the control rods are withdrawn, a fission chain reaction begins within the fuel elements. The water in a PWR is kept at a high pressure with a pressurizer, so it does not turn into steam. The hot water passes through a steam generator, where it passes its heat into cold water, which is then turned into steam. The steam is then run through a turbine to turn a generator and generate electricity. This steam then runs through a condenser, so its heat can be siphoned off by cool water. The reactor and the steam generator are kept within a fortified containment structure, which no radioactive materials ever leave.*

Reproduced courtesy of Alex Wellerstein (Stevens Institute of Technology, Hoboken, NJ, USA).

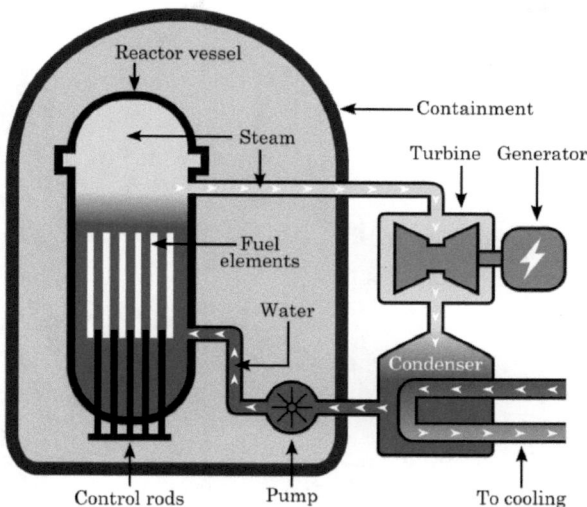

Figure 3.2 *A schematic representation (not to scale) of the basic principles of a boiling water reactor (BWR). Once the control rods are withdrawn, a fission chain reaction begins within the fuel elements. This reaction boils the water inside the reactor vessel producing steam. This steam is then passed through a turbine, which is connected to a generator that produces electricity. The steam is then passed through a condenser, which uses cold water to transform the steam back into cold water. The now-hot water from the condenser is pumped to a cooling tower, while the now-cold water inside the reactor loop is pumped back into the reactor. The reactor itself is kept within a fortified containment structure.*
Reproduced courtesy of Alex Wellerstein (Stevens Institute of Technology, Hoboken, NJ, USA).

force. In addition, longer-range conventional bombers were being developed. While this aircraft reactor experiment did not lead to use by its military sponsors, the program demonstrated the possibility of using a molten salt system that is currently an active generation IV design (see Chapter 17).

Also, Weinberg was the first to use a thorium fuel cycle in which neutron bombardment of thorium led to production of U-233, which can sustain a chain reaction. Such a reactor has reduced radioactive waste and enhanced safety features compared with U-235 designs. Through the years, thorium reactors have gained many adherents and something of a cult-like following with Alvin Weinberg cast as a hero.

In contrast with the aborted nuclear aircraft program, the quest for a nuclear-powered submarine, which was authorized at the end of 1947 led to testing of a prototype pressurized-water reactor (PWR) in 1953 and the launching of the Nautilus submarine in 1954 (Figure 3.3). The most notable voyage of the Nautilus was a trip that it completed under the North Pole. This success enabled Westinghouse to emerge as a leading developer of nuclear reactors. The Westinghouse PWR became one of two dominant designs, worldwide, for the next seven decades. The other leading design was General Electric's (GE) boiling-water reactor (BWR).

Figure 3.3 *Launch of the US Submarine* Nautilus *by Mrs. Mamie Eisenhower on January 21, 1954 at Groton, Connecticut.*
© Photo by Popperfoto via Getty Images/Getty Images.

The goal for the Naval nuclear submarine program of extended underwater performance was achieved with Nautilus staying submerged for up to two weeks in comparison with 1950's diesel vessels capacity of remaining submerged for only two days. Contemporary nuclear submarines can now remain under water for as long as four months.

Successful use of a nuclear reactor to power the Nautilus proved the viability of harnessing nuclear energy for civilian applications. However, there was little interest in developing reactors for production of electric energy either by the government or by industry. The government was concerned with the use of scarce enriched uranium for weapons and industry had little incentive to pursue nuclear energy when large supplies of inexpensive coal were available.

However, leaders in government felt pressure from political and international relations points of view. The legacy of Hiroshima and Nagasaki positioned the United States as the perpetrator of a horrendous act in the eyes of many throughout the world. Soon after these attacks on Japan, the United States initiated research studies of the effects of radiation on the survivors and those who were in utero, thus keeping the suffering of the bombing victims in public view. In Japan, survivors of the atomic bomb attacks were referred to as *hibakusha*. The *hibakusha* were treated as outcasts and considered by some to carry "the blood of the devil" https://www.newsweek.com/hiroshima-survivors-contagious-radiation-1522944.

Many Manhattan Project scientists and administrators continued to feel great angst during the post-war years. Notable among that group was Atomic Energy Commission (AEC) administrator David Lilienthal, who spoke frequently about the desirability of promoting peaceful uses of atomic energy. He expressed a desire to show that the "[mushroom] cloud had indeed a lining of silver."

Impact of Atmospheric Testing

Negative public perceptions of nuclear development were exacerbated by atmospheric testing of nuclear weapons at a site 65 miles north of Las Vegas, Nevada. The first detonation occurred on January 27, 1951. Atmospheric tests continued until 1963, when they were terminated by the Limited Test Ban Treaty. More than 100 atmospheric tests were conducted during that 12-year period.

Public concern about the radiation dangers continued to grow. Across the world these concerns intensified on March 1, 1954 after the first test of a thermonuclear weapon took place in the Marshall Islands and proved to be more than twice as powerful as anticipated. Lethal fallout descended on a Japanese fishing boat, the *Lucky Dragon*, and the resulting fallout exposure led to the death of one member of the crew six months later, as well as significant health consequences for the other 22 crew members. This event triggered worldwide protests against nuclear weapons and increases in fear about nuclear radiation. Negative attitudes toward nuclear weapons carried over to general opposition to nuclear energy.

In the United States, the government was secretly conducting studies of the radiation effects on human exposure to fallout and the atmospheric tests lead to independent civilian studies as well. In 1953 a particularly high radioactive fallout test resulted in the deaths of a quarter of the sheep in Utah and Nevada. Contrary to the evidence, the AEC claimed that the deaths were due to malnutrition.

As leading scientists began speaking out in opposition to atmospheric testing, a group in St. Louis, Missouri began a study of children's baby teeth. Between 1951 and 1970 more than 300,000 baby teeth were collected. Results showed that children born in 1957 had nine times more strontium-90 in their baby teeth than those born in 1951. This radioactive isotope was transmitted to children through digestion of milk produced by cows that had grazed on land contaminated by fallout. Strontium has chemical properties similar to calcium and consequently is readily absorbed by growing human bones. Strontium-90 has a half-life of 29 years, thus presenting a significant health hazard. While the impact of this study was not consequential until the 1960s and 1970s, its origin in the early 1950s is indicative of public concern.

While Lilienthal was initially a lone voice promoting nuclear energy, members of the US Congress and the Executive Branch increasingly sought civilian applications of nuclear technology to gain greater public acceptance of nuclear initiatives and as an action that would position the United States in a favorable light in the international arena. Toward that end, government leaders sought active participation by the power industry in developing nuclear reactors.

Industrial leaders saw nuclear energy as economically unfeasible. They also faced other impediments as they contemplated developing a nuclear reactor program. These included a limited nuclear engineering knowledge base, a wall of secrecy surrounding nuclear design information, and a government monopoly on enriched uranium supplies.

The AEC responded to these barriers by establishing a nuclear technology training school at Oak Ridge National Laboratory in 1950. They also introduced an industrial participation program in 1951 that provided access to classified nuclear reactor designs and invited industry players to submit proposals for development of their own designs.

Nuclear reactor design was inextricably linked to weapons development. The most direct connection was at Hanford, Washington, where reactors were used to produce plutonium for bombs (see Chapter 2). This connection between power reactors and weapons production not only justified secrecy, but also it encouraged multiple designs of nuclear reactors for dual use—power and weapons.

As the Soviet Union developed nuclear capability (demonstrated by the explosion of a fission bomb in 1949 and a thermonuclear device in 1952), it became a formidable rival to the United States on the world stage. In 1954, the Soviet reactor at Obninsk was connected to the civilian power grid, thus establishing a lead in the development of peaceful applications of nuclear fission.

Members of the US Congress and the Administration also took note that the United Kingdom was building a nuclear reactor for civilian use: the Caldor Hall reactor in Cumbria became operational in 1956.

Atoms for Peace

These various developments in the 1950s stimulated President Eisenhower to launch Atoms for Peace, a major initiative announced in his landmark speech before the United Nations General Assembly on December 8, 1953.

Eisenhower's speech was heard around the world. It dwelled on the inescapable dualism of atomic energy: its capacity for destruction as well as creation. Its sober tone nevertheless conveyed hope and, more concretely, a shift in US nuclear policy. Originally the United States kept its nuclear knowledge under lock and key—a strategy that had admittedly failed to prevent espionage even during the years of the Manhattan Project. Instead, the US government would seek to share its know-how with the world and Eisenhower proposed communicating through an international atomic energy agency, which would also confiscate and guard fissile material from governments around the world.

The event encouraged private industry participation in developing nuclear energy for commercial and civil uses. A key step in this transition was the passage of the Atomic Energy Act of 1954, which allowed the government to share technical information, previously held secret, with private organizations. The act also allowed patent processing for methods for atomic energy production.

Soon after the signing of the Atomic Energy Act of 1954, the new AEC announced its Five-Year Plan to support the commercialization of a variety of reactor concepts, including the Shippingport Pressurized-Water Reactor, Experimental Boiling-Water Reactor (EBWR), Sodium Reactor Experiment, Homogeneous Reactor Experiment-2 (HRE-2), and Experimental Breeder Reactor-II (EBR-II). It also established the Power Demonstration Reactor Program, which more thoroughly meshed private commercial and industrial capabilities with government-led nuclear research. Work undertaken on these programs provided many (often expensive) lessons that kindled a mixture of both hope and hesitation for the prospects of nuclear power.

With the strong governmental backing of nuclear energy, utilities previously wary of involving themselves in technology began to show interest. Within a year of the Atomic Energy Act, construction began on the first privately financed nuclear power reactor for the Commonwealth Edison company of Illinois, southwest of Chicago. In 1955, the utility contracted with GE to design, construct, and place in operation a 192-megawatt electric light-water BWR at a cost of $45 million. Construction began in 1956 and the reactor, known as Dresden 1, became operational on July 4, 1960. The utility was pleased with its operation, which continued until 1978. At a time when a bewildering array of experimental reactor designs presented an even wider array of unforeseen technical and operational difficulties, the relatively smooth running of Dresden 1 caught the attention of other utilities.

Commonwealth Edison was not alone among utilities in backing nuclear energy. The day after President Eisenhower signed the Atomic Energy Act, representatives of New England's major electric utility companies met to establish a framework that would implement a program to produce civilian nuclear energy. They outlined a plan to establish a generating company that would sell atomic energy to member companies; it became Yankee Atomic Electric. The AEC came forward with $5 million toward the estimated construction cost of $35 million with the owners of the parent company paying the additional amount. The AEC also agreed to waive the fuel costs for five years. The reactor design chosen was the Westinghouse 145-megawatt PWR, with Stone and Webster developing the other components of the power plant.

Construction of Yankee Rowe began in April 1958. Compared to later construction of nuclear power plants, the time for completion of this early reactor was exceedingly short: just three years. It was not only built in a timely fashion, with power being produced in 1961, but it was completed at a cost 23% below the original estimate. This engineering achievement merited high recognition at its inception and continued with operational success for 30 years, until questions about structural integrity led to its closing in 1992.

With both the USSR and the United Kingdom actively developing civilian applications of nuclear fission, Eisenhower and his senior advisors were concerned that the United States would lose out on opportunities to leverage its nuclear knowledge and experience in promoting relationships with non-aligned countries and uranium ore producing countries.

In his UN presentation, Eisenhower called for a de-escalation of weapons competition with the Soviets that included dismantling weapons and using the resulting fissionable materials for peaceful purposes. While this proposal struck a response chord and met with universal approval, it was unrealistic and was not pursued.

The International Atomic Energy Agency (IAEA)

Eisenhower called for an international organization to promote peaceful uses of nuclear energy; the result was the International Atomic Energy Agency (IAEA) located in Vienna, Austria. The IAEA became the world's foremost intergovernmental forum for scientific and technical cooperation for peaceful uses of nuclear energy. The Atomic Energy Act of 1954 enabled the United States to engage in domestic and foreign activities including the sharing of fissionable materials.

The United States entered into 39 bilateral agreements that promoted good will and a positive image on the world stage, including building research reactors, training researchers and technicians, and promoting applications for medicine, agriculture, and other fields. Agreements were established with various European and Asian countries, including the United Kingdom, France, Italy, Spain, Japan, the Republic of Korea, and Thailand. Research reactors were provided to 26 countries, of which eight, including India, Pakistan, Israel, and South Africa, initiated successful nuclear weapons programs.

After India exploded a nuclear bomb in 1974, there was great concern about the worldwide distribution of highly enriched, bomb-grade uranium by the United States as part of the Atoms for Peace program. As it had been exporting 1,500 lbs. of highly enriched uranium (HEU) per year (i.e., enough to make 15 Hiroshima bombs), the United States launched the Reduced Enrichment Research and Test Reactor program (RETR) to replace (HEU) in test reactors with low-enriched uranium.

Among the unintended consequences of the Atoms for Peace program was the establishment by the Soviet Union of a parallel program in countries with which it was aligned. These included China and North Korea, which later developed their own nuclear weapons.

With the United Kingdom and the Soviet Union programs to develop nuclear power for civilian use and the United States' Atoms for Peace initiative placing civilian applications of nuclear energy as a key component of its international agenda, developing nuclear energy for civilian use became a high priority. The Atoms for Peace program also positioned the United States in a dominant nuclear commercial position worldwide.

However, US government programs to encourage the power industry to develop power reactors failed. Not only was nuclear power unable to compete with the economics of coal, but the lack of expertise, the inherent dangers from fission technology, and uncertainty about the availability of fuel were all impediments to action by industry. However, the growing impatience of the White House, the US Congress, and the

Department of State with industrial inertia created concern among industrial leaders that Washington would move to nationalize nuclear energy and preempt the future involvement of private enterprise.

After failed attempts to engage industry, in the spring of 1954 the AEC issued a new program inviting industry to participate in the building and operations of a nuclear power plant. The leadership of the AEC also made clear to industrial leaders that action on their part would mean promoting a national agenda. The Atomic Energy Act of 1954 also provided further incentives that enabled corporations to own and operate reactors and to retain patent rights. Further, steps were taken to declassify nuclear technology.

In his diary, Eisenhower summed up government perspective on development of a nuclear power plant, writing that a plant would be a "symbol" of a "peaceful and purposeful" America. He continued that this action would "generate free world respect and support" for American foreign policy. Eisenhower and the National Security Council saw this as a factor in Cold War strategy by gaining influence and support from non-aligned countries.

Congress was debating whether nuclear power should be pursued as a public or private activity in this context. With mounting pressure on the private sector, Pennsylvania's Duquesne Light Company stepped up with an offer to develop a plant in Shippingport, Pennsylvania located about 40 miles from central Pittsburgh. While the cost of electricity from a nuclear plant was significantly greater than from a coal plant, the additional cost in Pittsburgh could be justified if the clean air of a nuclear plant was considered. Pittsburgh had been one of the most seriously polluted US cities, so limiting the use of coal in the region was quite welcome.

While a prototype plant was originally planned for Shippingport, national priorities led to construction of a fully operational facility using technology employed in the design of the Nautilus submarine and systems that had been studied and tested at Argonne National Laboratory. When the program to develop nuclear energy to power an aircraft carrier was canceled by Eisenhower, its plans for a basic reactor were shifted to Shippingport.

Shippingport Reactor

To expedite the building of the Shippingport reactor 90% of the cost was financed by the AEC, which assumed legal liability for any problems that might arise. Rickover oversaw the design of the reactor, with Westinghouse serving as the general contractor and the expertise of Argonne National Laboratory supporting the project.

Construction of this PWR started on Labor Day in 1954, with the reactor achieving criticality on December 2, 1957.

The reactor was originally designed for a naval ship using fuel that was enriched to 93% U-235 (compared with natural uranium, which only contains 0.7% U-235). This use of weapons-grade enriched uranium was a legacy of the naval program having originated with a mandate to construct a reactor for a submarine which necessarily required a compact source of energy, thus employing the most highly enriched uranium available. While the high level of uranium enrichment was used in nuclear submarines, it was understood that commercial reactors would operate at a low level of enrichment of just under 5%. While being used for civilian applications, the Shippingport reactor was not a commercial design, but rather a military design used by a civilian utility company in implementing an expedited timetable.

An innovation in the Shippingport reactor was the addition of a blanket of U-238 placed around the core for the purpose of producing plutonium when the U-238 absorbs a neutron. The plutonium can then act as fuel when it absorbs a neutron and undergoes fission. The reactor functioned in this configuration for seven years when the core was replaced with U-233 from government stockpiles and the blanket was replaced with natural thorium. The energy then originated from the fission of U-233 while more U-233 was produced when the thorium in the blanket absorbed a neutron.

The energy output of the Shippingport reactor was around 60 megawatts—an amount of energy capable of providing electricity for about 11,000 average homes. Over the next several years, the operating standard for nuclear reactors grew to a level of around 1000 megawatts. While a factor of 17 smaller than a more optimum power output level, the Shippingport reactor operated successfully as part of the Pittsburgh area grid for 25 years, providing proof of concept that led to the development of a robust commercial nuclear power industry. When the United States hosted the UN Atoms for Peace Conference in Geneva in 1958 it was able to point to the operation of the Shippingport reactor as an example of its commitment to civilian energy.

Development of Water-based Reactors

The 1950s saw the emergence of a new industry having multiple players, with construction companies eager to gain a foothold but dependent on private-sector utilities for contracts. The third player was the government, which funded and operated the national laboratories where the knowledge base for the industry resided.

Rickover was able to bring these entities together by overseeing a partnership of the Westinghouse Corporation with Argonne National Laboratory that provided services to Duquesne Electric Company. What emerged was a development track that established pressurized water as the dominant reactor design in the United States and elsewhere.

The Argonne Lab engaged with research and development staff from industry, including GE, whose engineer, Samuel Untermyer II, while working at Argonne Lab, developed the basic design for a BWR (see Figures 3.1 and 3.2).

The BWR was seen as a simplification by its developers who oversaw several iterations of the design through joint Argonne–GE collaboration. The success of the work at Argonne enabled GE to engage with Pacific Gas and Electric Company to embark on the construction of a BWR reactor at a location about 30 miles east of San Francisco, California. Work began at what became known as the Vallecitos Nuclear Center on January 1, 1956. When the reactor became operational on October 15, 1957, it was the first privately owned and operated reactor to deliver electricity to a public utility grid. The power output was 60 megawatts, that is, in the same ballpark as the Shippingport reactor.

While it produced significant amounts of electricity to the grid, it was known as a test reactor. When the Vallecitos reactor became operational, it replaced Argonne as the center for BWRs. It was used to explore a range of operational issues as well as serve as a non-governmental site for the training of operators and other support personnel. It had the distinction of receiving the AEC's first license: "Power Reactor No. 1."

The Vallecitos reactor provided reactor development and operating experience for GE, as it prepared to develop a significantly larger BWR in Illinois for the Dresden Nuclear Power Station located about 60 miles from Chicago. In 1955, the Commonwealth Edison company contracted with GE to construct a 192-megawatt electrical generator, with just over three times the output of the Vallecitos plant, only a year after the Atomic Energy Act allowed private companies to own and operate nuclear facilities. Due to the many uncertainties of entering into such a new enterprise, Commonwealth Edison partnered with a consortium of eight companies who between them shared one third of the cost of the contract. At the time, GE gained experience with a low-enriched reactor design at Argonne and was moving ahead with the Vallecitos plant. The only alternative contractor was Westinghouse, whose experience was with a HEU fuel that could not be used in a commercial reactor. Construction of the Dresden reactor began in May 1956 and the reactor became operational on July 4, 1960, three years after the completion of the Vallecitos reactor.

At about the same time as the Dresden plant was under construction, a PWR Westinghouse design (with about the same electrical power output) was being constructed in New England. The hesitancy of industry to adopt nuclear power is illustrated by the fact that ten power utilities shared the cost of the Yankee Rowe plant, where construction began in 1957, later achieving a self-sustaining nuclear reaction in August of 1960.

Thus, by 1960 both GE and Westinghouse had constructed commercial nuclear power plants, with GE building a BWR and Westinghouse building a PWR. While these two designs dominated the nuclear power plant construction field worldwide for decades, that development pattern did not take place rapidly in 1960. Westinghouse, GE, and the various utility companies found that their nuclear plant construction costs were higher than expected, and that related costs (e.g., insurance) were formidable. Furthermore, each had to contend with the issues of having trained personnel and a supportive local population. The manufacturing corporations and the utility companies did believe that, in the long run, nuclear power would play an important role in meeting the nation's energy needs and were therefore engaging in the enterprise sufficiently to forestall nationalization. However, the construction experiences with Dresden and Yankee Rowe failed to galvanize rapid expansion of new initiatives.

In conclusion, from the end of the Second World War to 1960, leadership in developing nuclear energy passed from government labs operating in secrecy and under military control to civilian organizations overseeing development and implementation by commercial companies operating in an open society.

Further Reading

The following suggestions explore the lives of the two most influential individuals in the development of nuclear reactor technology in the United States: Admiral Hyman G. Rickover and scientist Alvin M. Weinberg.

Francis Duncan, *Rickover and the Nuclear Navy: The Discipline of Technology*, Naval Institute Press, 1989.

A biography written by an associate of Rickover who recounts the development of the Navy's nuclear propulsion program and the domineering personality of Rickover, who enabled this development.

Alvin M. Weinberg, *The First Nuclear Era: The Life and Times of a Technological Fixer*, American Institute of Physics, 1997.

Weinberg's autobiographical account of the development of nuclear reactor developments, starting with the first reactor constructed by Fermi through the work of Weinberg at the Oak Ridge National Laboratory.

4

US Developments 1960–2000

To the critics, nuclear power is an incredibly complex and dangerous way to boil water. To the proponents, nuclear power is an ingenious and safe way to boil water, and one hell of a way to make money.

— Michio Kaku and Jennifer Trainer[1]

Commercial Reactor Development

Following the development of the Dresden 1 BWR and the Yankee Rowe pressurized-water reactors (PWR) in 1955 and 1956, emboldened utility companies began construction on ten additional nuclear reactors in the 1950s. This preceded an even larger boom in the 1960s and 1970s that resulted in the construction of almost all the reactors still operating in the United States in the mid-2020s. As regional energy utility companies became knowledgeable about nuclear energy and the economics appeared competitive with coal stations, there was considerable enthusiasm for development of nuclear power.

During the 1960s, construction was started on 60 nuclear power reactors in the United States, 23 of which have permanently closed, leaving 37 in operation in the mid-2020s. Building activity continued to surge during the 1970s with construction starting on an additional 62 reactors, 57 of which remain in operation in the mid-2020s.

These early reactors experienced numerous performance malfunctions during their years of operation, with the most serious being stuck valves and leaks that released radioactive contaminants into the surrounding ground water. Although regulations have since tightened, components advanced, and operations improved, such leaks have not been eliminated. In Illinois since 2007, there have been at least 35 reported leaks, spills, or other accidental releases of water contaminated with radioactive tritium. A leak discovered in 2007 at the Cordova Quad City boiling-water plant took eight months to plug and led to groundwater radiation readings up to 375 times the amount allowed under federal safe drinking water standards. In 2014, 500,000 gallons of high radioactive water were released from the Dresden Power Plant. Contamination was found in the plant's sewer line and miles away in an urban sewage treatment plant.

---

[1] Michio Kaku and Jennifer Trainer, *Nuclear Power: Both Sides: The Best Arguments for and Against the Most Controversial Technology.* W.W. Norton, 1983.

Nuclear Energy. Edward A. Friedman, Oxford University Press. © Edward A. Friedman (2025). DOI: 10.1093/9780198925811.003.0004

Almost all commercial nuclear energy in the United States has been implemented with the two basic designs—PWRs and boiling-water reactors (BWRs)—with Westinghouse and General Electric as the primary contractors. Although these designs overtook the commercial reactor market, several initiatives explored alternative designs. These included a liquid sodium-cooled reactor (see Chapter 19), a high-temperature gas-cooled reactor (see Chapter 21), and an organic fluid-cooled reactor.

The Infinite Fuel Reactor: The Promise and Disappointment of Fermi 1

Prior to the widespread search for uranium, fuel availability—later a strength of nuclear energy—was still uncertain. Eventually, engineers and physicists developed "breeder" reactors that would, once loaded with an initial fuel supply, produced their own additional fuel through neutron irradiation in the reactor core.

In the mid-1950s many scientists believed that development of breeder reactors would provide a limitless supply of reactor fuel and dominate the future commercial market for nuclear reactors. Plans moved forward for sodium-cooled breeder reactors as sodium, unlike water, does not absorb neutrons and supports a more efficient use of the available neutrons. The use of liquid sodium as a coolant instead of water supports an energy-producing chain reaction while making a surplus of neutrons available, since liquid sodium is a more efficient moderator than water. These surplus neutrons could then be absorbed by U-238 to produce Pu-239, which can also serve as fuel for fission.

However, Fermi 1, the first liquid sodium-cooled reactor, experienced a serious malfunction[2] that generated significant anti-nuclear sentiment in the United States. The construction of the 150-megawatt (MW) electric reactor began in 1956, soon after development was underway for Dresden 1 and Yankee Rowe. The Fermi 1 reactor attracted much attention because of the attractive possibility of developing a successful breeder reactor, thus guaranteeing a limitless supply of nuclear fuel. In the 1950s it was widely believed that the supply of natural uranium was limited and would run out if reactors were widely used. Therefore, there was great interest in developing a breeder reactor.

The road leading to the construction of Fermi 1 became a lesson in how not to pursue regulatory oversight. At the heart of the matter the conflict within the Atomic Energy Commission (AEC) of both regulating and promoting nuclear energy became blatantly obvious and set the stage for the eventual abandonment of the AEC structure in favor of the Nuclear Regulatory Commission in 1974.

The design of Fermi 1 was based upon a research reactor known as the Experimental Breeder Reactor-I (EBR-I) developed by a team from Argonne National Laboratory at a government facility that became known as the Idaho National Laboratory. EBR-I's purpose was to demonstrate the viability of the breeder design concept. Success was

---

[2] On October 5th, 1966, the Fermi 1 reactor suffered a partial meltdown due to the blockage of the liquid sodium coolant. This disruption caused many observers to question the viability of the liquid sodium cooled design.

achieved at a low level of power output of 200 kilowatts of electricity on August 24, 1951. On November 29, 1955, EBR-I experienced a partial meltdown when inadequate coolant fluid flow allowed the reactor to overheat. This event and its related questions about a possible loss of coolant appeared in a report to the AEC about potential problems for Fermi 1. At meetings that were called to evaluate a construction license for Fermi 1 there were calls to conduct additional research prior to embarking on implementation of construction plans for Fermi 1.

These AEC review committee reservations were reviewed at Congressional hearings where plans for additional testing were put forward. However, AEC chairman Louis Strauss expressed little concern for these calls for further testing and expressed determination to move forward with construction. Strauss strongly resisted Congressional calls for caution and refused to share the full content of the AEC review committee publicly. The AEC voted to issue a construction permit on August 4, 1956, with one commissioner voting no. This was done despite the analysis by the AEC review panel that recommended further study of the design.

Following the groundbreaking that took place on August 8, the powerful United Auto Workers (UAW) union brought a lawsuit to halt construction. At the same time, the AEC review committee's report was made public with the UAW's inflammatory interpretation of a report concern, noting, "In everyday language, this means the reactor might convert itself into a small-scale atomic bomb."

In the ensuing public debate, Detroit Power Company president Walter L. Cisler claimed that the issue was not simply safety, but rather the reflected hostility to private development of nuclear power. Claiming that the entire argument opposing Fermi 1 was politically motivated he asserted that "We are headed down the road to a socialist state," and that those speaking out against the design were " . . . prepared to use any subterfuge to keep atomic power development in the hands of the government."

This argument was countered by Michigan senator Patrick V. McNamara, who stated, "I reject the myth . . . that this is solely a fight between public and private power interests," and went on to declare that if "these questions raised solely by laymen—who knew little or nothing about the complexities and technicalities of atomic reactors? No. They were raised by the AEC's own Advisory Committee on Reactor Safeguards. And its questions to date have never been answered."

As this controversy played out, construction continued. Given the innovative nature of the design, it is not surprising that several problems emerged that required correction and modification of the original design. The construction issues, along with the legal controversy, led to delays that resulted in reactor startup in 1963—three years beyond the original target date.

The UAW lawsuit was heard in March of 1960. The following June, the verdict was handed down that the construction license was illegal and all work was required to stop within 15 days. This action led to an appeal to the US Supreme Court, which ruled in favor of continuing the construction of Fermi 1. Work continued and the AEC approved an operational license.

An initial chain reaction was successfully achieved on July 12, 1963, and the reactor began operations at a low level. The reactor was close to being able to deliver energy to the grid when on August 6, 1966, a partial meltdown occurred due to a metal plate blocking the flow of coolant. There was no external release of radioactivity, and the closed reactor was ready to be restarted in July 1970.

Repairs were implemented, but another mishap occurred when a pipe carrying sodium burst, leading to an explosion that caused further damage that was also repaired, but delayed operations to a time when a license extension was required. On August 27, 1972, the AEC issued a denial of application for license extension. In November this decision was accepted and shutdown was initiated.

The result of the legal case, as well as the accidents, had a significant negative impact on public perceptions of nuclear energy. While the accidents were not close to having a major environmental impact, the press portrayed them as being close to catastrophic. John G. Fuller's *We Almost Lost Detroit* appeared soon after the shutdown and implied that the design was inherently unsafe; Fuller opined that Detroit was not destroyed only by a matter of luck. The Fermi 1 partial meltdown also inspired folk singer Gil Scott-Heron to write a song also called "We Almost Lost Detroit," which gained wide popularity. Its ominous lyrics raised the specter of nuclear annihilation and likened the reactor to a ticking time bomb. The psychological impact of this catchy song helped shape negative public perceptions of nuclear power. The Fermi 1 accident contributed, in a major fashion, to the anti-nuclear movement in the United States. Rumination about the event reverberated for many years despite Fermi 1's fully functional safety systems and no release of radioactivity.

Clinch River Reactor

When the Fermi 1 reactor was nearing the end of its existence, the AEC was promoting a large-scale (350-MW electric) sodium-cooled breeder reactor to be built at Clinch River, Tennessee, just outside of Oak Ridge. Both the public and President Jimmy Carter opposed the plan. In addition to the dismal results from the Fermi 1 experience, the plant's initial cost was high and continued to increase substantially the year after it was introduced. There was also fear that the production of plutonium in the breeding process might result in a threat of nuclear weapons production.

Congress pushed for the development of Clinch River despite Carter's opposition. He responded in a 1979 speech and said that the proposal ". . . marches our nuclear policy in exactly the wrong direction. It is no time to change America into a plutonium society."

President Regan opted to continue the project, which was finally shuttered by the US Senate on October 26, 1983.

New Reactor Initiatives

Development of nuclear reactors picked up pace through the 1960s and the early 1970s. While it continued to evoke polarized opinions, optimism for the technology by the federal government and utility companies, coupled with rapidly growing energy demands, drove dozens of announcements for new nuclear reactors each year. In 1973, publications by the AEC forecasted the need for around 1,200,000 MW of nuclear power (~ 1000 reactors) by the year 2000. Two years later, in his State of the Union address President Gerald Ford announced the more moderate but still ambitious goal of having "200 major nuclear power plants" operating within 10 years.

However, unrest in the sector continued to grow. A wave of regulation hit utility companies hard. The 1970s saw the establishment of the Environmental Protection Agency (1970), the Occupational Safety and Health Administration (1973), and the Nuclear Regulatory Commission (1975), as well as other federal and state agencies. The portion of the US GNP under regulation grew from 8.2% to 23.7% between 1965 and 1975. Price regulations had caused utility companies little trouble through the 1960s as labor and material costs fell. However, the energy shocks of the early 1970s and the accompanying inflation created an entirely different atmosphere. Fuel prices increased by 160%, which produced a ripple effect throughout the economy. Commodity prices rose between 60% and 70% and labor costs rose 40% between 1970 and 1976.

Utility companies were initially fearful to launch rate cases asking for substantial increases in consumer prices. When rate increases were finally implemented, public utilities commissions often became jammed with outlandish backlogs. It wasn't until the mid-1970s that utility-regulated rates increased by more than just 2–4%.

The resulting utility "death spiral" stalled new large capital spending. As the regulated rates that utility companies were allowed to charge customers lagged below their new costs, profits dropped. Investor returns fell to within just 2% of what they could receive from interest on loans, making it difficult to raise money by issuing equity. Instead, utility companies took on enough debt to trigger regulated debt-to-equity ratio limits that further restricted their ability to raise funds. Spending on system maintenance and new-generation technology declined, with multi-billion-dollar nuclear stations being among the first to fall by the wayside.

While the energy crisis prompted awareness for the increased need for domestic energy production that could be met with nuclear solutions, it was an extremely difficult time to build expensive, long-term projects.

Nuclear energy was becoming increasingly expensive in response to persistent litigation from anti-nuclear environmental groups. For example, concern for thermal pollution in waterways led utilities to gradually and reluctantly build expensive and imposing cooling towers to disperse heat into the atmosphere.

A sweeping and onerous safety principle emerged, that of ALARA ("As Low as Reasonably Achievable"), in addressing nuclear waste, radiation, and accident risk. This led to substantial equipment and compliance costs for safety and radiation protection regulation. ALARA created effectively limitless justification for new expenses, as any profit margins became funds that could "reasonably" be spent toward marginal increase in safety.

ALARA significantly influenced regulations developed both by the Atomic Energy Commission (1946–1974) and its successor, the Nuclear Regulatory Commission (1974–present).

Pro-nuclear and anti-nuclear voices became entrenched in a battle to win over public opinion, with groups volleying petitions back and forth and appealing to legislators on both sides of the debate. The anti-nuclear voices were also adapting to the new circumstances. Compared to the early days, when radiation from weapons testing stoked resistance to nuclear energy, new environmentalists had adopted fairly sophisticated critiques of large, centralized power stations and bureaucratic utilities. On top of financial and regulatory hurdles, fervent opposition to new nuclear plants in New York and California gave utility companies a taste of the political and civil challenges of pushing these projects through.

The opposing forces around nuclear energy created a roller coaster for the sector: over 100,000 MW of new reactors were announced in the first half of the 1970s, only to be canceled over the remainder of the decade and into the first half of the 1980s.

Organic Nuclear Reactor

In the early days of commercial nuclear reactor development there was serious interest, both in the United States and in Canada, in the use of organic fluids for cooling of nuclear reactors. Organic fluids, which are widely used, are those that contain carbon as a chemical component. They are present in paints, varnishes, adhesives, and in degreasing and cleaning agents, and in the production of dyes, polymers, plastics, textiles, printing inks, agricultural products, and pharmaceuticals. A major attraction for organic fluids is their being corrosion-resistant, unlike water, which, when used as a reactor coolant, can only be used on expensive corrosion-resistant metals. The use of organic cooling fluids in a reactor also avoided the danger of steam explosions in the event of the reactor overheating. However, a major drawback in using organic fluids is that they have a much lower capacity to transport heat and require more powerful pumps to sustain higher liquid flow velocities than in water-cooled reactors.

Another disadvantage to using organic fluids as a reactor coolant is that high-energy neutrons in the reactor core causes breakdown of hydrogen–carbon chemical bonds, which releases free hydrogen and creates a thick tar-like component in the cooling fluid. This process, known as polymerization, proved to be too difficult an obstacle to overcome in designing organic fluid-cooled nuclear reactors.

The AEC implemented a demonstration project to explore the feasibility of organic fluid-cooled nuclear reactors and built a plant in Piqua, Ohio in 1963, which was closed in 1966. During the operation of this reactor problems developed with the control rods and fouling of the cooling surfaces, neither of which could be overcome through the modifications that were attempted. Although sometimes discussed, to date there has been no continuing program to realize the use of organic fluids as coolants for nuclear reactors.

Megatons to Megawatts Program

At the time of the breakup of the Soviet Union, the country experienced an economic crisis and underwent a serious rethinking of its nuclear position. In an Op-Ed published in the *New York Times* on October 24, 1991, MIT physicist Thomas Neff proposed an innovative plan that would provide the Russian Federation with liquid resources while meeting nuclear fuel needs of the United States. The following year the United States and the Russian Federation formalized a contractual agreement for 500 metric tons of highly enriched uranium (HEU) from Russian nuclear warheads to be converted to a 5% enrichment level for use in American reactors. This agreement was planned as a 20-year program. The dilution from a 90% enrichment level down to the 5% reactor fuel level was performed in the Russian Federation.

This program ran according to schedule between 1993 and 2013, resulting in the delivery of 15,000 tons of reactor fuel. This was the largest and most successful non-proliferation program ever implemented in the history of nuclear weapons treaties. During this 20-year period as much as 10% of the electricity produced in the United States was generated by fuel fabricated from this Russian source. However, the long-term consequences of this arrangement resulted in the Russian Federation becoming the dominant supplier of HEU. By 2022, Russia was the source for approximately half of the world's HEU and about one quarter of the nuclear fuel used in the United States. Russian nuclear fuel was exempt from the trade embargo imposed on Russia following its invasion of Ukraine. Additionally, Russia was the only source of 19% HEU required for the operation of Bill Gates's Terrapower reactor—viewed as an important component of the next-generation nuclear reactor initiative to combat global warming. In response, the United States initiated new efforts for domestic production of HEU.

Further Reading

"Megatons to Megawatts Program," *Wikipedia* (last modified July 20, 2024). https://en.wikipedia.org/wiki/Megatons_to_Megawatts_Program
A discussion of the United States–Russia Highly Enriched Uranium Purchase Agreement and its impact on Russia becoming the dominant supplier of enriched uranium.
"Nuclear Power in the USA," *World Nuclear Association* (last modified August 27, 2024). https://world-nuclear.org/information-library/country-profiles/countries-t-z/usa-nuclear-power.aspx
This website provides a detailed history of the development of nuclear energy in the United States.

5

Nuclear Resistance

Oh, meltdown. It's one of those annoying buzzwords. We prefer to call it an
"unrequested fission surplus."

—Mr Burns, *The Simpsons*[1]

Opposition to Nuclear Energy Emerges

As development of civil nuclear power advanced in the 1960s, opposition emerged in
step. This opposition pushed a worldview of fear and indignation toward all things
nuclear, mixing reactions to the use of nuclear weapons, the impact of atmospheric
nuclear tests, the possibility of nuclear reactor accidents, and a general fear of radiation.

The Lingering Shadow of Hiroshima and Nagasaki

The horrific nature of the atomic explosions at Hiroshima and Nagasaki is only partially
conveyed by the number of deaths that occurred from the two detonations. While there
is a large uncertainty surrounding that total, figures range from 110,000 to 210,000
deaths, which places the consequences of nuclear weapons into a grim context.

When the Second World War ended, studies began to observe, track, and treat the
thousands of casualties suffering from radiation-caused infirmities. Around 200,000
people survived the bomb blasts but were instead exposed to radiation and experienced
trauma because of the bombings. Survivors also experienced psychological injury arising from the extreme stress induced at the time of the attacks and the fear of the potential
long-term consequence of radiation illness. Among the survivors were a group of 3,289
who were in utero at the time of the bombings. The suffering, maladies, and premature
deaths among the survivors were a constant reminder to the world of the dangers of
nuclear technology. It became impossible to divorce any discussion of nuclear fission
from the pall generated by the detonations of August 1945. Studies and news of the
after-effects continued for more than 60 years. By 2002, 853 excess deaths from solid

---

[1] Mr. Burns, *The Simpsons*, S3E5: "Homer Defined," Fox, October 17, 1991.

Nuclear Energy. Edward A. Friedman, Oxford University Press. © Edward A. Friedman (2025). DOI: 10.1093/9780198925811.003.0005

cancers had been identified, along with 98 deaths due to radiation-induced leukemia. These and related studies were a continuing reminder both of the bombings and the dangers of radiation.

The Era of Atmospheric Nuclear Testing

Soon after the conclusion of the Second World War, the United States initiated a program of atmospheric nuclear weapons testing. In July 1946, Operation Crossroads in the Bikini Atoll of the Marshall Islands was the site of the first postwar test series.

Discussion and planning for these tests began within weeks of the surrender of Japan. The tests aimed to study the nuclear weapon damage to naval ships. A fleet of 95 ships was then assembled for these tests, with many of the ships having been captured from Japan in the closing days of the war. There was a support fleet of more than 150 ships used to conduct these tests. Between 1947 and 1958, 23 nuclear detonations shook Bikini Atoll.

Animals were placed on these ships to observe radiation effects, with the objective of evaluating the diagnosis, treatment, and general protection of possible future ships' crews that may be exposed to nuclear weapons. However, Operation Crossroads provided only 15% of the participating personnel with film-badge dosimeters, designed to measure their potential exposure to radiation. Secrecy imposed on these tests prevented many veterans from communicating knowledge of potential harmful effects of radiation to their civilian doctors. The US Department of Defense belatedly began studying the potential exposure of radiation among the military and civilians who participated in these tests. In early 1978, the Department of Defense organized a Nuclear Test Personnel Review to identify those who had participated in the atmospheric nuclear weapons test program and to determine the extent of their exposure to ionizing radiation.

Ten years later, the US Congress passed a bill to provide additional compensation to "atomic veterans" who developed specific types of cancer due to exposure to radiation. In 1994, President Bill Clinton assembled an Advisory Committee on Human Radiation Experiments, and then issued a formal apology to the victims of these experiments. In 1996, Congress repealed the Nuclear Radiation and Secrecy Agreement Act, removing the last barrier to the victims to speak out freely about what had occurred.

Potential harm caused by radioactive fallout from nuclear weapons tests was a constant concern from the very first nuclear explosion at the Trinity test on July 15, 1945. The removal of civilian populations from the fallout region in New Mexico was incomplete, leading to years of negative health consequences for residents who were not evacuated. The ensuing controversies led to the 1990 Radiation Exposure Compensation Act, which provided payments to those who developed health conditions that could plausibly be attributed to fallout from weapons tests.

The United States conducted nearly 200 atmospheric nuclear weapons tests between 1945 and 1962. About 100 of these tests took place in Nevada and a nearly equal number in the Pacific Ocean, mainly in the Marshall Islands. Further, starting in 1949, the

Soviet Union conducted 219 atmospheric, underwater, and space tests that paralleled the American program. A vast area of Kazakhstan was used for nuclear testing by the Soviet Union.

The United Kingdom, France, China, India, Pakistan, and North Korea also joined in atmospheric nuclear testing, bringing the global total of tests to 520. The United Kingdom conducted 45 atmospheric tests in Australia and the Montebello Islands. France conducted tests in Algeria and in French Polynesia. In 1963, the Limited Test Ban Treaty was negotiated by the United States, the United Kingdom, and the Soviet Union. This ended most atmospheric tests, although France, China, India, and North Korea did not sign.

Populations around the world were aware of atmospheric testing. With the highly visible impact of nuclear weapons on Hiroshima and Nagasaki, it was not surprising that anti-nuclear movements began to proliferate.

The *Lucky Dragon* Incident: A Catalyst for Nuclear Fear

The most visible and dramatic nuclear test—that of an early version of a thermonuclear device—occurred on March 1, 1954. Thermonuclear energy is derived from the fusion of light elements, rather than from the fission of heavier elements. The Sun's energy is produced by such reactions.

The fusion reaction used in the Castle Bravo test involved the element deuterium fusing with the element tritium. Both deuterium and tritium are isotopes of hydrogen. They each contain one proton in the nucleus, but deuterium has an additional neutron and tritium has two additional neutrons (Figure 5.1). The result of deuterium fusing with tritium is a helium nucleus plus an energetic free neutron (Figure 5.2).

The fusion reaction is triggered by a fission explosion that not only supports the high impact collisions between the tritium and deuterium but also generates the tritium component of that reaction. The tritium is produced from solid lithium deuteride when neutrons from the fission explosion are absorbed by the lithium that is present, causing it to split into an alpha particle (the nucleus of the helium atom) plus tritium. Lithium is defined by having three protons in its nucleus. It has two stable isotopes, one with three neutrons known as lithium-6, and one with four neutrons known as lithium-7

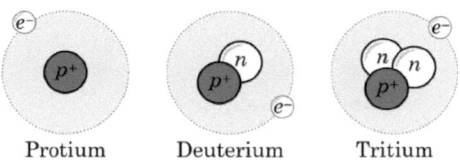

Protium   Deuterium   Tritium

Figure 5 1 *The three main types of hydrogen.*
Reproduced courtesy of Alex Wellerstein (Stevens Institute of Technology, Hoboken, NJ, USA).

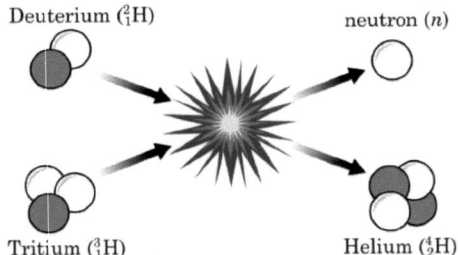

Figure 5.2 *Deuterium–tritium fusion yielding helium and a neutron.*
Reproduced courtesy of Alex Wellerstein (Stevens Institute of Technology, Hoboken, NJ, USA).

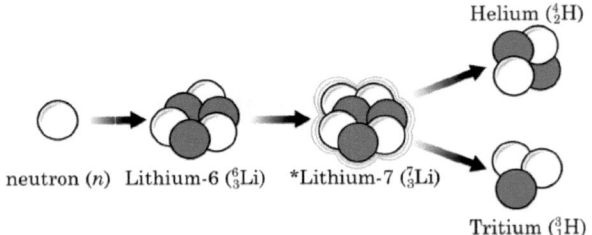

Figure 5.3 *Lithium-6 capturing a neutron to produce lithium-7, which splits into helium and tritium.*
Reproduced courtesy of Alex Wellerstein (Stevens Institute of Technology, Hoboken, NJ, USA).

(Figure 5.3). Lithium-7 is far more abundant, constituting 95% of naturally occurring lithium, with lithium-6 found in the remaining 5%. Thus, a fission explosion triggers the production of tritium from lithium, which in turn combines with the deuterium nucleus to provide the energy. It is this energy that is used in a thermonuclear bomb.

Those who designed the Castle Bravo test were most familiar with the transformation of lithium-6 into tritium plus an alpha particle when bombarded with a neutron. Their expectation was that this would provide the required amount of lithium for the fusion reaction. However, they were unaware that, for high-energy neutrons, lithium-7 also yielded tritium plus an alpha particle with a single neutron emission (Figure 5.4).

This added source of tritium increased the yield of the Castle Bravo test from an anticipated equivalent of six megatons of TNT to an actual yield of 15 megatons—an increase by a factor of 2.5!

$$n + {}^{6}_{1}\text{Li} \rightarrow {}^{4}_{2}\text{He} + {}^{3}_{1}\text{H}$$
$$n + {}^{7}_{1}\text{Li} \rightarrow {}^{4}_{2}\text{He} + {}^{3}_{1}\text{H} + n$$

Figure 5.4 *Lithium-6 and lithium-7 reactions with neutrons.*
Reproduced courtesy of Alex Wellerstein (Stevens Institute of Technology, Hoboken, NJ, USA).

The Castle Bravo explosion was about a thousand times more powerful than the bomb dropped on Hiroshima and stands out as the sixth-largest nuclear explosion in history. The unexpected magnitude of this blast had a devastating impact on the devices placed to observe its detonation. Many of the permanent buildings on the far side of the atoll, some of which contained devices to monitor the event, were severely damaged. Little of the desired diagnostic data on the impact of the explosion was collected. Many of the instruments designed to transmit their data back before being destroyed by the blast were vaporized instantly, as were most of the instruments that were expected to be recovered for future analysis.

There were about 20,000 people living on the nearby atolls of Rongelap and Rongerik, both of which experienced fallout along with 13 additional islands and atolls in the region. This affected population was evacuated 48 hours after the detonation. Despite this large-scale emergency evacuation, over $43 million was eventually awarded to 1,200 islanders who experienced various illnesses due to the fallout.

A Japanese fishing boat, the *Lucky Dragon No. 5*, was 86 miles from the Bikini Atoll test site. Its location would have been a safe distance away from a six-megaton explosion, but in reality it was squarely in the middle of a region of intense fallout from the actual blast. Members of the crew witnessed a false "sunrise" that lit up the western sky, engulfing the boat in a bright light. The surprised fishermen knew the light came from unnatural causes, and nine minutes after seeing the unexpected glow, a roaring sound passed over the boat.

As the 23 crew members pulled in their fish-laden nets, strange clouds formed in the direction of the blast sound. This was followed by white rain that coated the ship and the men with a gritty ash that covered their bodies and entered their mouths and eyes. It turned a dark blue tuna on the deck a ghostly gray.

Rain and ash fell on the ship for five hours. Many members of the crew reacted with dizziness, vomiting, and fevers. They were covered in, swallowed, and inhaled highly radioactive coral remains incinerated by the explosion. The *Lucky Dragon* returned to port two weeks later, at which time most of the crew exhibited headaches, bleeding gums, skin burns, and hair falling out in clumps. They were all hospitalized.

The impact of the *Lucky Dragon* event reverberated throughout Japan and elicited consternation worldwide. The arrival of the crew, who were all suffering from acute radiation disease, brought the American testing program, which had been going on in the Pacific for eight years without attracting much notice, to the attention of the press and the public.

The shock and horror of the radioactive contamination of the crew of the *Lucky Dragon* triggered anti-nuclear movements that started in Japan and spread throughout the world. Here was material evidence that turned the invisible threat of radiation to plain images of contaminated ash and afflicted fishermen.

The *Lucky Dragon* event triggered a strong reaction in Japan with enduring consequences. In 1954, following wide-scale reports of the *Lucky Dragon* episode, Japanese film company Toho released a horror film in which a fictional prehistoric reptilian (kaiju-monster) Godzilla is awakened and powered by nuclear radiation (Figure 5.5). In addition to the death and destruction wrought by Godzilla, the film contained political and social undertones relevant to Japan at that time.

Figure 5.5 *Original Godzilla monster (1954).*
Reproduced from Toho Company Ltd. (1954). Wikimedia Commons. https://commons.wikimedia.org/wiki/
File:Godzilla_(1954).jpg.

The impact of *Lucky Dragon No. 5* was enormous and was followed by 33 subsequent films from Toho, and five films based on the same theme were produced in the United States. *Godzilla Minus One,* the most recent version of the Godzilla story, continues to relate the consequences of radiation to an international audience; it was awarded the Oscar for best visual effects at the 2024 Academy Awards!

In addition to films, the Godzilla theme entered the cultural arena with television series, novels, comic books, video games, and other merchandise. Godzilla is familiar to people throughout the world. Regardless of the evidence assembled to support the benign impact of radiation from nuclear reactors, the story of Godzilla continues to suggest otherwise.

Godzilla is not the only superhuman creature empowered by exposure to nuclear radiation. Spider-Man, the Hulk, and others are also part of the genre that has captured the public's imagination.

After Castle Bravo

After the Castle Bravo test, there was immediate pressure on the American government to be more forthcoming regarding its ongoing testing program. In particular, *Operation Ivy*, a film about a prior test with the same name, had been made for internal government use, but public lobbying brought about its widespread release. It was seen widely in the United States and around the world. This film brought to viewers the enormity of thermonuclear weapons that were vividly seen to be a thousand times as powerful as the fission weapons dropped on Hiroshima and Nagasaki.

The film contained staggering illustrations of this power. For example, it showed that if such a weapon were detonated at the Capitol Building in Washington, DC, everything would be annihilated for three miles in all directions. While there was a general understanding of the destructive capacity of nuclear weapons from widely distributed documentation of the aftermath at Hiroshima and Nagasaki, the enormity of the vastly more destructive capability of thermonuclear weapons, demonstrated by Operation Ivy, triggered a tidal wave of anti-nuclear activism.

A notable example of that activism were public pronouncements of the Greater St. Louis Committee on Nuclear Information, an organization composed of Washington University scientists who called for a halt to atmospheric testing. This group was instrumental in organizing a study of baby teeth that had absorbed the radioactive isotope strontium-90 that occurred in fallout from nuclear testing. Teams were organized to collect baby teeth from school children in the St. Louis region. They set out to collect 50,000 teeth a year and ultimately assembled over 320,000 teeth.

They found a striking correlation between the occurrence of strontium-90 and testing activity. They showed that children born in 1963 had levels of strontium-90 that were 50 times higher than those in children born in 1950, prior to the advent of large-scale atmospheric testing.

These results strongly influenced US president John F. Kennedy to pursue the Partial Nuclear Test Ban Treaty with the United Kingdom and the Soviet Union. The treaty was signed in August of 1963.

Opposition to Reactors

The early 1960s saw nuclear power development accelerate in the United States, with plants being planned and financed for a growth surge that lasted until the 1980s, when the economics for fossil fuel energy generation became advantageous and the impact of the Three Mile Island accident in 1979 (see Chapter 11) took hold. The surge in nuclear energy development provided new soil for an anti-nuclear movement that had its roots in weapons testing. Nuclear energy not only came closer to home, but nuclear safety was now under the purview of for-profit monopoly utility companies—who many trusted even less than the government.

Ravenswood Reactor

A particularly egregious example of unrealistic overreach by a utility company occurred in New York City, when Consolidated Edison (ConEd) sought to build the world's largest nuclear power plant at Long Island City's Ravenswood site just across the East River from the heart of the city. In October of 1962, two months after ConEd commissioned a plant along the Hudson River at Indian Point, 30 miles north of New York City, it submitted a proposal to the Atomic Energy Commission (AEC) for a 1000-megawatt reactor to be built less than two miles from Times Square. The proposed plant had about four times the capacity of the Indian Point plant.

Their application to the AEC for a plant that would have been the world's largest, with a power output that would have exceeded all of the existing nuclear power plants combined, was submitted in early December.

In hindsight, this was an extravagant proposal, but at the time ConEd was not alone in considering construction of nuclear reactors in urban areas. The proposal for the Ravenswood location in Queens became the subject of debate and discussion in New York City in 1963.

In retrospect, it is astonishing that such a proposal was put forward, but at the time, though it raised eyebrows, it was accepted as plausible. The proposal had an impact on those living in the neighborhood of the proposed plant that went beyond raised eyebrows. A community meeting was held in February 1963 in this middle- and working-class neighborhood, where the local civic leader captured the mood of attendees with the declaration that he was opposed to the project and would continue to be opposed until convinced otherwise.

In contrast, the ConEd leadership demonstrated blasé unawareness of the depth of public opposition. ConEd chairman Harold C. Forbes testified at a Congressional hearing in April that ". . . one or two people have raised some questions about the genetic effects of radiation and so forth." He said that such concerns were "rather silly" and went on to render a summary opinion that ". . . the public in general has reached the point where it has accepted nuclear plants as a matter of course."

In contrast to Forbes's insouciant views about placing a large nuclear reactor in the center of the nation's largest city, there was forceful critical testimony at the same Congressional hearing by former AEC chairman David E. Lilienthal, who stated "I would not dream of living in the borough of Queens if there were a large atomic power plant in the region, because there is an alternative—a conventional thermal power plant as to which there are no risks."

The debate over the pros and cons of having a large nuclear reactor in an urban location continued throughout 1963 with ConEd arguing for the cost effectiveness of eliminating the need for expensive transmission lines and the positive health benefits of power that did not cause air pollution. Opponents of nuclear power tended to focus on the biological effects of radiation.

The Ravenswood proposal proved to be a stimulus for public engagement with nuclear issues. New York City community forums, civic organizations, and politicians grappled with these issues. In response to the need for accurate, unbiased information, an independent group of academics came together at Rockefeller University under the leadership of Rene Dubos. There, faculty and graduate students from the New York Metropolitan area participated in the Scientists Committee for Public Information (SCPI), whose members adhered to a policy of presenting information to the public that met the objectivity and accuracy standards of peer-reviewed academic publications. SCPI was asked to present its information to organizations including the Kiwanis Club, the Rotary Club, and political and community groups. The scope of topics addressed by SCPI expanded to include the advisability of fallout shelter construction and the nuclear arms race.

Reevaluation of power plant guidelines at the AEC also took place with a move toward more stringent constraints on urban development. The result of these reviews led ConEd to withdraw its application for a reactor at Ravenswood in early January 1974. It accepted the expense of building transmission lines and developed a plan to acquire hydroelectric power from Canada.

Today, the Ravenswood Generating Station is an operational oil- and natural gas-fired power plant whose red and white-striped smokestacks can be seen when crossing the Queensboro Bridge.

Shoreham Nuclear Power Plant

The general acceptance of nuclear power in the 1960s led the Long Island Lighting Company, in 1965, to propose building an 820-megawatt plant in Shoreham, Long Island, a sparsely populated location situated 70 miles from the center of New York City (Figure 5.6). The site, located on the water of Long Island Sound across from Connecticut, was five miles north of Brookhaven National Laboratory, which housed several research reactors.

Construction of General Electric boiling water reactors at the Shoreham location began in 1972, with the facility being commissioned in 1986. The planning and construction of these reactors proceeded without undue concern from the local population until the 1979 Three Mile Island accident sparked a demonstration of 15,000 protesters opposing the facility. This outpouring of public opposition was the largest event of its type in the region's history. The intensity of the protest led to the arrest of 600 people, many of whom attempted to scale the fences surrounding the construction site.

The Three Mile Island incident (see Chapter 11) prompted the Nuclear Regulatory Commission to require operators of nuclear power plants to adopt evacuation plans in cooperation with state and local governments. This requirement empowered on groups opposing the Shoreham plant, including the Sierra Club, the Audubon Society, and other environmental organizations. The physical location of the Shoreham plant, on a relatively narrow island located 70 miles from New York City, presented

Figure 5.6 *The Shoreham nuclear power plant.*

Reproduced from Paul Searing (2007). Wikimedia Commons. https://commons.wikimedia.org/wiki/File: Shoreham_Nuclear_Power_Plant.jpg. under a Creative Commons Attribution 2.0 Generic (CC BY 2.0).

serious challenges to the development of an evacuation plan. Organizers mobilized many opposition groups and exerted considerable pressure on political leaders.

On February 17, 1983, the local legislature voted overwhelmingly in favor of a resolution stating that the region could not be safely evacuated in the event of an accident at Shoreham. This led to the governor of New York ordering that state officials should not approve any evacuation plan.

After the plant's completion in 1984, permission was received to operate it at a low power level. However, local authorities kept up opposition to the endorsement of any evacuation plan, thus keeping the potential for operations in limbo. The Chernobyl disaster occurred on April 26, 1986, which sealed the fate of Shoreham. As its prospects waned, the Long Island Lighting Company continued its advocacy for the plant until finally ending its efforts on May 19, 1989. The termination of this initiative included an agreement which added a 3% surcharge on the electric bills of Long Island residents for 30 years to pay off the $6 billion cost of the unused plant.

Energy Crisis of 1973 and Nuclear Consequences

In 1973, Egypt and Syria attacked Israel in what was a failed effort to regain land that Israel had annexed several years earlier. In retaliation for the intervention of the United States on the side of Israel in that conflict, Arab states, as part of the Organization of Oil Exporting Countries (OPEC) since 1960, initiated an embargo on oil shipments to the United States. OPEC founders included Kuwait and Saudi Arabia as well as Iran, Iraq, and Venezuela.

In response to the oil embargo and the resulting fuel crisis, President Richard Nixon proposed Project Independence to ensure US energy independence by 1980 with substantial expansion of nuclear energy by 2000. Initial reactions to proposals for expanded use of nuclear energy in 1973 were met with acceptance, if not enthusiasm. However, as nuclear reactor plans evolved, so did organized programs, whose aim was to eliminate nuclear energy from the available options.

Among the most effective anti-nuclear organizations to emerge during this period was the Clamshell Alliance, founded in 1976 to oppose the Seabrook Station Nuclear Power Plant in New Hampshire.

First proposed in 1966 by the Public Service Company of New Hampshire (PSNH), the Seabrook Station was planned to consist of two large nuclear reactors, each with a capacity of 1244 megawatts electric. Construction was impeded because of regulatory issues, protests, and lawsuits. It took a full 14 years before a single plant was commissioned for operation in 1990; the second reactor was abandoned, resulting in a loss of $800 million and the bankruptcy of PSNH in 1988—then the fourth-largest bankruptcy in US history.

The Clamshell Alliance was a well-organized initiative that first adhered to non-violence principles. There were multiple sit-ins at the construction site with the largest taking place on May 1, 1977, with over 2,000 protesters of which 1,414 were arrested. The authorities imposed bail of $500, which those arrested refused to pay. After two weeks, the prohibitive cost of keeping the protesters incarcerated led to the release of 550 demonstrators without requiring payment of bail. At the time, it was one of the largest mass arrests in the history of the United States. Clamshell Alliance organizers used this detention for training and networking, which strengthened the organization and resulted in it becoming a model for demonstrations held elsewhere, particularly in California.

Nuclear Fear

The opposition to nuclear development regarding the Ravenswood, Shoreham, and Seabrook reactors all dealt with explicit reactor proposals that could be discussed and debated in concrete terms. However, there are negative attitudes toward nuclear energy in society triggered by discussions of public policy unrelated to specific proposals. These are reactions to the very concept of atomic energy. This negativity has become embedded in society's collective psyche due to multiple factors. Contributing to impact has been the devastation at Hiroshima and Nagasaki, the effects of radiation on those on board the *Lucky Dragon No. 5*, the deaths that resulted from the Chernobyl disaster, and other nuclear accidents that, though not directly lethal, caused widespread panic. These grim examples of nuclear dangers have been the subject of numerous news accounts, books, films, and public presentations. In many cases the material presented to the public is exaggerated and sensationalized. A notable example is the Netflix production *Chernobyl*, in which firefighters who experienced lethal doses of radiation responding to a catastrophic accident at a Soviet nuclear plant are portrayed as being able to transmit radiation sickness by being in proximity with others. Additionally, there have been many science fantasy horror movies that attribute terrifying consequences to radiation exposure.

In his book *The Rise of Nuclear Fear*, Spencer Weart notes that a "risk is especially likely to be feared if it is largely unknown—not only invisible but something new, outside normal experience, mysterious. A risk also seems magnified if it has elements evoking dread: something bearing a stigma of horrid associations, something with a potential for unbounded catastrophe." He adds to this analysis the heightened discomfort of a risk to which the public feels they did not consent, or where risk mitigation is the responsibility of an organization seen as untrustworthy. These and other factors well established by social scientists apply to nuclear energy.

Weart reports on studies that correlate opposition to nuclear energy with various demographic factors. These studies find a predilection for nuclear energy among engineers and hard scientists, with opposition being found more strongly among artists, social scientists, and those engaged in humanistic pursuits. Differences in public policy attitudes are also found between individuals who tend to systematically study issues and those who glean their beliefs from news items, movies, and other social media.

Through the years, a major shift has occurred in these cultural and social divisions in the United States as members of the Democratic party (who were previously deeply opposed to nuclear energy in the 1990s) are now proponents of nuclear energy. Especially striking is the role of John Kerry, who led the campaign to close the Integral Fast Reactor Program at Argonne National Laboratory in 1993, but who emerged as spokesman for the United States in promoting a tripling of nuclear power at COP28 in 2023.

Shifts are also visible in polls of the general population where sentiments in favor of nuclear energy have increased in recent years. This trend seems to correlate with growing concern about global warming and fading memories of Chernobyl and Fukushima.

Nuclear Protests and Extremism

While circumstances differ across projects and countries have diverse contexts when promoting nuclear power, there are many similar elements to fears about nuclear power. Whatever the project and wherever its location, opposition to a nuclear energy initiative attracted passionate participants. In France 1975–1977, some 175,000 people protested nuclear power in ten demonstrations, while in West Germany 1975–1979, some 280,000 people protested in seven demonstrations. Similar events took place in other countries around the world, with particularly large protests in the months following the accidents at Three Mile Island, Chernobyl, and Fukushima (discussed in Chapters 11, 12, and 13, respectively).

In the period 1970–2020 there were 91 attacks on nuclear facilities and nuclear scientists that resulted in 19 fatalities and 117 injuries. Incidents took place in 25 countries with the largest number in France and Spain (14 events in each country), followed by Japan (nine events) and Germany and the United States (seven each). None of the attacks resulted in radioactive fallout or environmental contamination, and most occurred outside a nuclear power plant. The largest number (29) took place between 1971 and 1980, and the lowest incidence was between 2001 and 2010, rising to 18 between 2011 and 2020. This extreme reaction seems to be an ongoing phenomenon and correlated with the level of nuclear development in society.

The Superphénix Breeder Reactor

Of the many nuclear power plants that attracted protests one that stands out as having a particularly consequential and complex history: opposition to the Superphénix breeder reactor at Creys-Malville, France. Perhaps the most bizarre and dangerous of all antinuclear episodes, this section first reviews some relevant history.

In the 1960s there was a widespread belief that the world's uranium supply would fail to provide the quantities needed to fuel the market demand for nuclear energy. This fear of a shortfall in nuclear fuel prompted interest in breeder reactors that could replenish the uranium fuel supply. In a breeder reactor neutrons emerge at high velocity and are not slowed down through collisions with a moderator. Neutrons thus can be absorbed by U-238 and transformed into plutonium, which can then absorb additional neutrons and undergo fission as well as create an excess supply of Pu-239. The plutonium produced can then serve as fuel for a nuclear power plant.

In 1962 France embarked on a fast neutron breeder development program in which liquid sodium was used to cool the reactor. This first such reactor was an experimental device operating at a power level of only 20 megawatts of thermal energy without electricity. This reactor, known as Rapsodie, operated successfully from 1967 until 1983.

In 1968, after Rapsodie had been operating a year, construction began on a larger 233-megawatt breeder reactor called Phénix, which went critical in 1973. Phénix was fueled with Pu-239 and cooled with liquid sodium. Since it sustained a chain reaction with fast neutrons that emerged from the fission reaction there was no need for a moderator. Phénix produced 233 megawatts of electrical energy and excess neutrons were absorbed in a uranium blanket surrounding the reactor core. These uranium atoms that absorbed neutrons were transformed into plutonium. For every 100 plutonium atoms that provided neutrons for the generation of heat, 116 plutonium atoms were created in the uranium blanket. Thus, the breeder reactor produced more nuclear fuel than it consumed. This was an extremely attractive outcome that could sustain the energy needs of France and others who pursued this technological wonder.

Phénix was connected to the electrical grid in 1973 and remained operational until 2009. Since it required a refueling every two months, which required shutting down, the reactor was operational only 65% of the time. However, it served its purpose of providing a learning environment for breeder technology that included a demonstration of using a fast reactor to burn nuclear waste. This was possible since the most troublesome radioactive elements in waste are those with heavy elements with long half-lives that could undergo fission when bombarded with fast neutrons. It was perhaps the location of Phénix in central-southern France, 83 miles from Marseille, that kept it from becoming the focus of large-scale protests. However, a reactor that drew its inspiration from the success of Phénix, the Superphénix, became a magnet for protests and controversy.

In 1973, the year of the OPEC oil embargo, Phénix was connected to the grid. While France was not directly targeted by the embargo, it nevertheless had a strong impact on France. French Prime Minister Pierre Mesmer initiated energy conservation measures and France saw the success of Phénix as providing a route to energy independence.

While the promise of energy abundance through the development of fast breeder reactors had been discussed for several years, in 1974 France, West Germany, and Italy jointly established the European Fast Reactor Corporation (NERSA), a limited international company, to pursue development of a 1,242-megawatt electric fast breeder reactor. Shares in NERSA were divided, France holding a 51% share, Germany 33%, and Italy 16%. New laws were passed enabling this international entity to implement a nuclear construction project.

As the site chosen for Superphénix, Creys-Malville, France is located 270 miles from the Italian border and 85 miles from Geneva, Switzerland. The designated reactor site was only 48 miles from the French city of Lyon and 58 miles from a major research center in Grenoble, France. As an international program that required legislation in three countries and which was within driving distance to two adjacent countries, the development of the Superphénix reactor attracted considerable attention.

When a public review of the proposed reactor's safety was conducted, 657 observations hostile to the project were received with over 2,000 signatures. Leading scientists from nearby Lyon and Switzerland's European Organization for Nuclear Research (CERN) were openly opposed to the project. The controversy's geographic impact grew to encompass a sizable portion of continental Europe. Of particular concern to those opposed was the inclusion of plutonium both as a fuel and as an output of the proposed reactor. Plutonium was the fuel for the bomb dropped on Nagasaki and as a free-standing substance it is extremely toxic. The radiation of alpha particles from plutonium is like that emitted by radium—an element with an unsettling history of causing radiation poisoning in the 1920s and 1930s. Also, that an isotope of plutonium, Pu-239, has a half-life of more than 24,000 years, added to its ominous reputation.

Despite the opposition, permits were obtained to build the Superphénix reactor and construction activities began in 1976. In July 1977, 30,000 protesters from France,

West Germany, Belgium, Switzerland, and Scandinavia marched on the Creys-Malville site where they encountered 5,000 riot police. The police used tear gas to disperse the marchers. Both police and protesters were injured; one police officer lost a hand when a gas canister that he was holding exploded, and one protester was killed when he was trampled by others fleeing from the tear gas. This unfortunate event failed in its purpose and was the last such mobilization to oppose Superphénix.

In October 1978, a group of 30 Swiss intellectuals and elected officials formulated a petition known as the Geneva Appeal opposing the construction of the Superphénix reactor. The petition was circulated throughout Europe and gained 50,000 signatures and stimulated multiple groups to mount campaigns to halt construction.

The construction work on Superphénix was not deterred by the many protests and Europe-wide opposition. Between 1976 and when the reactor was commissioned in 1986, Phénix, which had all the same characteristics as Superphénix, continued to operate without protest and, in fact, maintained unchallenged operations until it was decommissioned in 2010. The difference in public response to the two reactors is probably due to the location of Superphénix, which was close to the borders of Switzerland and Italy and which required international approvals, thus making it a topic of immediate interest to a wide European audience. Also, its proximity to Lyon and Grenoble (and their concentrations of scientists) as well as to CERN (populated by leading international physicists) meant the presence of many anti-nuclear activists. While scientists and engineers, as a segment of society, were generally supportive of nuclear energy development, specialists in physical science were often outspoken opponents. Many in this technical community had developed negative attitudes triggered by the centrality of plutonium in the design of this breeder reactor.

Among the lay population in the region lived Chaim Nissim, an engineering graduate of the École Polytechnique de Lausanne who had emigrated to Switzerland from Israel with his family. He was born in 1949 and participated in the march on Creys-Malville in 1977 at the age of 28.

An account of the activities of Chaim Nissim is included here because he illustrates the extremes that emerged in society resulting from an antipathy to nuclear reactor development. Paradoxically, he was a pacifist who engaged in the use of explosives, and an associate of violent terrorists while he held a role in civic government.

He wrote in his blog that he believed that the reactor being built at Creys-Malville could potentially experience a nuclear explosion and that all means should be taken to prevent its being built, including non-violent illegal means. He participated in dynamiting the electrical pylons that supplied the construction site. Nissim held that the use of explosives was consistent with pacifist ideology provided the explosives were used against physical targets, rather than against humans.

As a pacifist activist who espoused non-violent protest, it is incredible that he contacted members of the Baader-Meinhof Group, a West German far-left militant group who were notoriously violent. They were the most militant extremist guerrilla group in Europe in the 1970s and 1980s, committed political murders, and participated in the airplane hijacking that led to the Israeli commando raid on Uganda's Entebbe Airport, in which 102 Israeli hostages were rescued.

Nissim traveled to Brussels where he met with the terrorists and obtained five Russian-made anti-tank rockets and a rocket launcher. The weapon was an RPG-7 portable reusable, unguided, shoulder-launched, anti-tank rocket launcher. It had been used by the Soviet Army since 1961 and was also used by the Irish Republican Army in Northern Ireland from 1969 until 2005 against British Army armored personnel carriers. It was designed to pierce armor and considered suitable for penetrating the barrier being constructed to house Superphénix. Nissim received instructions on the use of this weapon from a Russian member of the group. This weapon was used to carry out the extraordinary attack on the Superphénix in 1982.

He rationalized this rocket attack as non-violent since the action was directed at a physical structure and not at construction workers. He and his collaborators attacked at night, believing that there were no workers present, from a hill in the countryside near the reactor site. It was later revealed that there was one person in the vicinity, who fortunately escaped harm. It was also fortunate that the facility had not yet been supplied with nuclear fuel and only suffered minor damage to the enclosure.

Nissim fled the location without being apprehended and eluded authorities, who tried for many years to track down the perpetrators of the attack.

Superphénix started its operation in 1986, four months before the Chernobyl disaster. After one year of operation a sodium leak occurred, which required changing the fuel handling system. This event triggered a new series of protests and lawsuits that tried to stop further activity at the reactor. The reactor was approved to restart, but in 1990 an air leak endangered the sodium coolant and forced it to again shut down.

Superphénix did not obtain approval to restart until 1994. However, during the shutdown, in 1992, authorities had decided to repurpose the reactor as a research facility. Renewed protests saw the formation of an organization known as Europeans against Superphénix, which obtained 22,500 signatures on a petition calling for the complete shutdown of the plant. Additionally, in April 1994, 300 activists walked from Creys-Malville to Paris to bring the issue forward on the national scene.

After another five months of operation, an argon leak occurred causing a shutdown. Reconsideration evaluations in the mid-1990s on the future of Superphénix looked quite different than it did in the mid-1970s, when the planning for its construction began. Twenty years later there was not the same urgency regarding the availability of nuclear fuel and the many problems encountered in operating Superphénix sapped much of the enthusiasm for the project. Also, in the 1990s numerous lawsuits came forward challenging the licensing of the design modifications. Finally, in June 1997, newly elected French Prime Minister Lionel Jospin announced the permanent closure of Superphénix. This was a political decision that ended the operations of the world's largest breeder reactor.

In 1985, Chaim Nissim was elected to the Grand Council of Geneva as a representative of the Green Party of Switzerland. He held that position until 2001. Twenty years after the rocket attack, the French statute of limitations protected him from persecution for that act. In his 2004 memoir *L'amour et le monstre: roquettes contre Creys-Malville*, he wrote that "We failed, as the closest rocket missed the important part that we targeted by one meter. It was nevertheless quite beautiful. And symbolically it was a token contribution to the larger movement."

Upon this public admission of his leadership role in the rocket attack he was expelled from the Green Party but was readmitted in 2006. Nissim's account is an extreme example of anti-nuclear fear in a citizen who is both an exemplar of a responsible citizen as well as an extremist engaged in violent rebellion.

While Chaim Nissim is clearly an outlier in the spectrum of citizen response to nuclear energy, he embodies the fears and ambiguities that have been pervasive in nuclear energy discourse for more than 60 years and which continue to be evident in the ongoing dialogue.

Perspective on Nuclear Resistance

Antipathy to nuclear energy has emerged from fear of radiation, the destruction of Hiroshima and Nagasaki, and years of nuclear testing, as well as nuclear reactor accidents and the specter of nuclear war. There is no obvious dominant promoter of nuclear antipathy, since all these factors contribute. There exists entanglement among factors, but the impact of each changes with time. An issue's prominence in the day's news influences its level of potency in generating fear.

There is something about nuclear antipathy that runs deep in the human psyche and allows the public to ignore millions of deaths from fossil fuel-generated air pollution. Instead, it encourages many to focus on the supposed dangers of nuclear waste, which has been relatively benign in its impact on human mortality.

Note that the protests and anti-nuclear energy events discussed here, for the most part, took place prior to the accidents at Three Mile Island, Chernobyl, and Fukushima. These three accidents were not the source of nuclear antipathy, but rather events that allowed deep-seated fears to become realities.

Further Reading

Chaïm Nissim, *L'amour et le monstre—Roquettes contre Creysmalville*, Favre, 2004.
This memoir, written by Chaim Nissim, provides unusual insights into the emotions and motivation of his unusual rocket attack on the Creys-Malville reactor.
David Ropeik, "How the Unlucky Lucky Dragon Birthed an Era of Nuclear Fear," *Bulletin of the Atomic Scientists*, February 28, 2019.
An account of the dynamics that led to the deep distrust of nuclear energy resulting from the Lucky Dragon No. 5 incident.
Spencer R. Weart, *The Rise of Nuclear Fear*, Harvard University Press, 2012.
This analysis of nuclear fear is a comprehensive and well-structured review of the human reactions that have evolved in response to the introduction of nuclear energy into society. His commentary on psychological factors that have affected attitudes toward nuclear energy are particularly insightful.

6

Reactor Development in Britain

It is with pride that I now open Calder Hall, Britain's first atomic power station.

— Queen Elizabeth II, October 17, 1956

UK Developments 1945–2000

Soon after the discovery of fission by Otto Hahn and Fritz Strassmann in Berlin in 1938, scientists in the United Kingdom began studies to evaluate the feasibility of a bomb initiated by fission. Rudolf Peierls and Otto Frisch completed the MAUD report in 1941, which described their calculations that a critical mass of ten kilograms of uranium-235 (U-235) would be large enough to produce an enormous explosion. They realized that a bomb of that size could be loaded on an existing aircraft and be deployed within two years.

In 1941 the conclusions of the MAUD report were communicated to Vannevar Bush and James Conant, who were actively considering nuclear weapons development in the United States. This led to a close working relationship in the Manhattan Project, where US and UK scientists collaborated on developing the bombs used in 1945 in Japan.

With the end of the war, the United States moved to disassociate itself from nuclear-related collaboration with the United Kingdom. This change in policy became apparent to the British leadership as the war was ending. Then Prime Minister Clement Atlee established a cabinet-level committee in August of 1945 to examine the feasibility of Britain embarking upon an independent nuclear weapons program. The US Atomic Energy Act of 1946 ended the American policy of sharing "restricted data" with its allies—which accelerated the British bomb initiative. Planning moved forward during the ensuing months, culminating in a formal decision in January 1947 to proceed with the development of a national atomic bomb program (Figure 6.1).

Given their past participation in the Manhattan Project, British scientists were intimately familiar with the options that were open to them to develop a bomb. The alternative of developing a bomb using U-235 versus plutonium-239 (Pu-239) was evaluated. The high cost of building a gaseous diffusion plant for the enrichment of U-235 was seen as prohibitive, given the post-war weakness of the British economy. Plans therefore settled on construction of reactors that could provide Pu-239 from uranium-238 (U-238) neutron absorption.

Nuclear Energy. Edward A. Friedman, Oxford University Press. © Edward A. Friedman (2025). DOI: 10.1093/9780198925811.003.0006

Figure 6.1 *Traveling Atom Train Exhibition (1947–1948) in which British scientists shared nuclear knowledge with the British public.*
By Courtesy of the University of Liverpool Library (Reference D974/1).

Windscale Pile Reactors

The design selected was based upon the X-10 Graphite Reactor built at Oak Ridge National Laboratory during the Manhattan Project. The X-10 was the world's second man-made nuclear reactor. The first was built under the direction of Enrico Fermi in Chicago in 1942 and had demonstrated the feasibility of a self-sustaining nuclear chain reaction. The X-10 was built for continuous operation for the purpose of producing plutonium.

The British design used natural uranium embedded in pure graphite that was air cooled. The site that was chosen was on the northwest coast of England. Two reactors were built, known as Windscale Pile No. 1, which became operational in October 1950, and Windscale Pile No. 2, which became operational in June 1951. These reactors were built with the primary objective of producing plutonium for bombs, although the construction provided for use of some of the energy by the local electric power utility.

However this was a small fraction of the 180-megawatt thermal energy output from this reactor.

Windscale Pile No. 1 was approximately 40 feet high and 24 feet wide. A total of 3,440 fuel channels were positioned horizontally through the reactor and each channel contained 21 foot-long aluminum fuel canisters. Canisters were pushed into the channels where they contributed to the fission process, thereby providing an opportunity for some of their U-238 contents to absorb a neutron and be transformed into Pu-239. The exposure time was limited to minimize the additional neutron absorption that produced plutonium-240 (Pu-240), since it interfered with the dynamics of a Pu-239 chain reaction. The reactor was cooled by air that was forced through the system by arrays of fans and sucked out of the reactor through a 400-foot chimney.

At an appropriate time, canisters would be expelled through the channel opening at the rear of the reactor, where they would fall into a container and be held in water pending retrieval at a later time. The canister would be held prior to being processed for isolation of plutonium.

This design was problematic since the canisters were not pushed through the channels in a precise fashion. Canisters that were pushed too vigorously would not land in the container holding cooling water but instead would land on the surrounding ground and split open. Manual handling needed to be implemented to relocate intact canisters and to clean up broken canisters. Occasionally, fuel canisters became stuck in the channels and burst open. It is not surprising that radioactivity was found around the site and in the nearby village. The British government kept this unfortunate release of radioactivity hidden from the public.

The Windscale piles were first-of-kind developments that presented operational difficulties not previously encountered. In particular, the graphite matrix that held the uranium canisters was subject to structural instabilities caused by neutron bombardment. The neutrons that flowed through the reactor introduced pockets of stored energy in the graphite with the associated potential of causing a fire if the energy was suddenly released. Experience with the graphite matrix led to initiating a process of forcing the release of the built-up energy by heating the graphite to a temperature exceeding 482 °F (250 °C). This annealing process was carried out periodically, although over time, changes in the structure of the graphite limited the effectiveness of the procedure.

On October 7, 1957, the reactor was subject to its ninth annealing, which resulted in a cartridge bursting and causing a fire in one of the channels. Nearly 48 hours elapsed before this fire on the discharge face of the reactor was discovered, and operators did not have a clear plan for dealing with the fire. They first tried to blow the flames out by running the fans at maximum speed, which resulted in an intensification of the blaze. They then tried to push the flaming cartridges out of their channels, which proved to be impossible.

The next attempt to quell the fire was to smother it with carbon dioxide gas. It happened that the nearby Calder Hall reactors, which were designed to be cooled with carbon dioxide gas, had just received a delivery of liquid carbon dioxide. Unfortunately, operators were unable to put a flow of carbon dioxide gas into place as needed.

Four days passed, during which there was significant release of radioactivity in the smoke and flames from the fire. The shielding surrounding the reactor became subject to intense heat of 2,400 °F (1,300 °C), which threatened structural collapse. The operators were forced to use water to quell the flames. This was a solution that the operators were reluctant to use since it was highly dangerous, as contact between water and the molten metal of the cartridges could potentially release explosive hydrogen from the water. Nevertheless, although water was cautiously injected into the flaming channels, this effort also failed!

As a last resort, the operators shut off all cooling and ventilating air entering the reactor. Fortunately, this act of starving the conflagration of oxygen proved successful and the flames gradually subsided. Water was then used to cool the reactor with the result that it became contaminated with radioactivity, which then spread to the area around the reactor.

While the filters that had been installed on the output of the smoke stack were able to contain most of the particulate matter from spreading beyond the location of the reactor, there was release into the atmosphere of radioactive material, which spread across the United Kingdom and Europe, including the highly dangerous polonium-210. This radioactivity was kept secret, along with many of the details of this accident. However, there was sufficient release of iodine-131 to impose restrictions on the consumption of milk in an area of 200 square miles surrounding the site of the reactor. This was to safeguard children whose thyroid glands were particularly vulnerable to damage from the absorption of iodine-131.

The Windscale Pile No. 1 accident stands as the third most serious nuclear reactor accident on record following Chernobyl and Fukushima. It is estimated that radioactivity released by Windscale caused approximately 100 premature cancer deaths. Windscale Pile No. 2 was closed and decommissioned soon after these events as continued operations were judged to be too dangerous. In hindsight, the way these reactors were operated was seen as quite flawed. The action of the British government to keep details of the accident secret precipitated a rupture in nuclear collaborations with the United States.

Magnox Reactors

The Windscale Pile No. 1 reactor design using air as a coolant was modified in the construction of the next generation of British reactors, which used carbon dioxide gas as a coolant, thus avoiding oxidation of the graphite matrix. These second-generation British reactors known as Magnox reactors came online in 1956 at Calder Hall, not far from the Windscale location (Figure 6.2).

The Magnox reactor design, as with the Windscale Pile reactors, was primarily optimized for production of plutonium, with power generation seen as a secondary benefit. In 1957, the British government decided to promote electrical energy generation by nuclear power despite the cost being 50% higher than coal power stations. This was justified by the co-production of plutonium and the leverage that having an alternative to coal generation provided when negotiating demands of the coal miners' union. A total

Figure 6.2 *Queen Elizabeth II visits Calder Hall for its ceremonial opening in 1956.*
Reproduced courtesy of the UK Nuclear Decommissioning Authority.

of 11 Magnox power stations were built containing 26 reactors. In addition, one reactor was exported to Italy and another to Japan. The first Magnox reactor, which was opened by Queen Elizabeth II on October 17, 1956, was closed on March 31, 2003, after 47 years of use. The last Magnox reactor was shut down in 2015.

Advanced Gas-cooled Reactor (AGR)

A next-generation gas-cooled reactor designed for more efficient production of electricity was introduced in 1962 but did not come online until 1976. A total of 14 Advanced Gas-cooled Reactors (AGRs) were built at six sites between 1976 and 1988.

The AGR was designed such that the final steam configuration was identical to that of conventional coal-fired power station. Thus, the same design could be used for the electric generator equipment.

The fuel for the AGR was uranium dioxide pellets enriched to 2.5–3.5%. The AGRs were designed to be refueled without shutting down. This on-load refueling was an important part of the economic case for the selection of this design. However, fuel rod mobility problems led to the curtailment of on-load refueling.

The AGR proved to be complex and difficult to construct. Notoriously bad labor relations at the time added to the problems. The lead station, Dungeness B, was ordered in 1965, with a target completion date of 1970. Thirteen years later, after problems with nearly every aspect of the reactor design, it finally began generating electricity in 1983.

In 2006 AGRs made the news when, under the Freedom of Information Act, *The Guardian* obtained documents claiming that the construction management team had been unaware of the extent of the cracking of the graphite bricks in the cores of the reactors. Further statements suggested that they did not know why the cracking had occurred and that they were unable to monitor the cores without first shutting down the reactors. This problem was acknowledged but not resolved.

Cracking of graphite bricks in AGRs places a time limit on the viability of an AGR. There are multiple causes of cracking that are difficult to control, including radiation damage to the structural integrity of the bricks as well as uneven heating and oxidation. AGR reactors therefore need to be monitored with regular inspections.

The challenges of operating AGRs built with structures composed of graphite bricks proved formidable. Since enriched uranium had become available in the international marketplace along with the development of international reactor design standardization initiatives, the UK developed interest in acquiring a pressurized water reactor (PWR). The only PWR to be built in the United Kingdom in the twentieth century was contracted in 1987 and commissioned in 1995 as Sizewell B.

The Sizewell B reactor became an option as a result of the Standardized Nuclear Unit Power Plant System (SNUPPS) initiative involving a group of American power industry companies, with Bechtel Power as the lead architect engineers and Westinghouse as the lead nuclear generator provider. Westinghouse sought to provide sufficient standardization for plants designed for different locations in order for Bechtel to be able to implement operations with minimal variations from site to site. The turbine generators were provided by General Electric.

Sizewell B used the SNUPPS design but with turbine generators that were like those used with the AGR reactors. A British construction contractor oversaw fabrication of the power station and EDF Energy, a wholly owned French corporation, was responsible for operating the facility. The parent company was Electricité de France, a state-owned corporation.

Sizewell B, Britain's only PWR, has been extremely successful. As with other PWRs, it has maintained an 18-month operating cycle, at or near 100% output during that time, followed by a one-month shutdown for maintenance and refueling. Built with an anticipated 40-year lifetime would result in shutdown in 2035. However, EDF plans to seek a 20-year extension for operation on the reactor until 2055.

Further Reading

Lorna Arnold, *Windscale 1957: Anatomy of a Nuclear Accident*, Palgrave Macmillan, 1992.
An official history of the Windscale accident that was commissioned by the United Kingdom Atomic Energy Commission.

"Nuclear Power in the United Kingdom," *World Nuclear Association* (last modified December 6, 2024). https://world-nuclear.org/information-library/country-profiles/countries-t-z/united-kingdom.aspx
A summary overview of the history of nuclear development in the UK.

Simon Taylor, *The Fall and Rise of Nuclear Power in Britain: A History*, UIT Cambridge, 2016.
A book recounting the history of reactor development in the UK that is written for a broad audience.

7

Reactor Development in France

In France, we have all kinds of things, we have the best cuisine in the world, a powerful industry, pétanque, a glorious history, we also have a privileged geographical location, the Eiffel Tower and angling, yes in France we have all that and much more. Yet we are missing one thing, an essential thing: oil. Oil we are forced to buy from others, expensive, too expensive. [. . .] In France, we don't have oil, but we have ideas.

—a TV advert written by a team at LINTAS, an ad agency hired by AEE[1]

Early Nuclear Developments in France

France has a long and involved history when it comes to nuclear physics. In 1896, Henri Becquerel discovered the radioactive behavior of uranium in his Paris laboratory. Just two years later, Pierre and Marie Curie announced their discovery of the radioactive elements polonium and radium, achievements that would earn them Nobel prizes. This pioneering work continued via their daughter Irene Joliot-Curie and her husband Frederic who, in 1934, discovered induced radioactivity, where a stable element becomes radioactive when exposed to a radioactive source. They too were awarded a Nobel Prize for this breakthrough.

Pierre Joliot-Curie was active in public affairs. In 1945 he was appointed the first High Commissioner for Atomic Energy (CAE) by French President Charles de Gaulle. In that capacity, Joliot-Curie oversaw in 1948 the construction of the first French atomic reactor. During that period, he was also an active member of the French Communist party, which he had joined in 1942. He was a leader in the party and became a founding member and president of the World Peace Council, where he advocated for nuclear disarmament to prevent atomic war.

On March 15, 1950, the World Peace Council, with Joliot-Curie as spokesperson, issued the Stockholm Appeal, which called for an absolute ban on nuclear weapons.

---

[1] Yves Bouvier. Du civisme économique à la citoyenneté globale: mises en discours et mises en images des politiques d'économie et de maîtrise de l'énergie depuis 1973, pp. 10–25.

Nuclear Energy. Edward A. Friedman, Oxford University Press. © Edward A. Friedman (2025). DOI: 10.1093/9780198925811.003.0007

The Appeal was signed by millions of people from around the world, including American composer Leonard Bernstein, artist Marc Chagall, French politician Jacques Chirac, American novelist Dashiell Hammett, author Thomas Mann, artist Pablo Picasso, playwright George Bernard Shaw, and composer Dmitri Shostakovich. In recognition for this work, he was the recipient of the first Stalin Peace Prize that was awarded in Moscow on April 6, 1951.

Joliot-Curie's role in the Communist party along with other leading French nuclear scientists reinforced the isolation of the United States atomic weapons program from its allies. Upon Joliot-Curie's promulgation of the Stockholm Appeal, he was dismissed from his position as head of the Commissariat a l'Énergie Atomique (CEA) in 1950. He was succeeded by physicist Francis Perrin, who had not signed the Stockholm Appeal.

The First Nuclear Reactors

Plans for the development of nuclear reactors were approved by the National Assembly in July 1952. Planning began at that time for the construction of three initial reactors along the same lines as the first reactors that were built in the United Kingdom and where similarly, reactor development was initially motivated by a desire to obtain a nuclear weapons capability. Although the French designs were like those of the British, they were developed independently. The first nine reactors built in France were known as UNGG (uranium naturel graphite gaz) units.

The first UNGG reactor had a power output of a mere two megawatts (MW) of electricity and the next two had outputs of 40 MW of electricity. They came online in 1956, 1959, and 1960, with the dual objective of producing electrical energy and plutonium.

The decision to develop an atomic bomb was made by the French government under President Pierre Mendes at the end of 1954. The use of nuclear weapons as a deterrent became national policy under Charles de Gaulle, who served as President from 1959 until 1969. The plutonium produced by the UNGG reactors fueled the first French nuclear test explosion that took place on February 13, 1960, in Algeria.

Construction for the last gas-cooled reactor built in France began in 1965 at Bugey with a power level of 555 MW electric. This reactor went online in 1972 and contributed to the grid until it closed in 1994.

The Messmer Plan

International events forced the French government to confront its dependence upon imported oil. The 1973 oil embargo imposed by OPEC caused acute French vulnerability to high oil prices. On March 6, 1974, Prime Minister Pierre Messmer announced a hugely ambitious nuclear power program aimed at generating most of

France's electricity from nuclear power. At that time, most of France's electricity was generated by foreign oil.

The Messmer Plan was adopted without public or parliamentary discussion or debate. The plan called for construction of around 80 nuclear plants by 1985 and 170 plants by 2000. While these projections proved to be overly ambitious, France did succeed in constructing 56 plants over the next 15 years, which was quite remarkable! Most noteworthy is the fact that the Messmer Plan led to France becoming the leader in nuclear power use among the developed nations of the world, with 70% of its electrical energy derived from nuclear fission.

Prior to the final gas-cooled reactor being built in France, there was an initiative, started in 1960, to use tested pressurized water reactors (PRWs). France collaborated with Belgium that year to build a unit that was designed by Westinghouse. Then in 1962, construction began at Brennilis on an experimental reactor that used natural uranium cooled with carbon dioxide gas and moderated with heavy water. This reactor, which became operational in 1967, had a heat generating capability of 70 MW.

In August 1975, two explosions at the site slightly damaged a turbine. This destruction was attributed to the Breton Liberation Front, a terrorist organization, and led to the closing of the reactor on an accelerated schedule.

As the Messmer Plan became actualized, construction began primarily with designs obtained from Westinghouse. The French company Framatome was aligned with Westinghouse until 1981, when it terminated the relationship and became the lead contractor for development of new nuclear reactors. The Framatome reactors in the 1970s generated around 900 MW electric, while in the 1980s this capacity increased to a maximum of 1500 MW.

The extraordinary vigor with which the Messmer Plan was implemented was a testimony to the French decision-making process. While other Western democracies engaged in intense public discussion and debate regarding nuclear power, France did not. The French have traditionally left decisions on science and technology policy to the government in power. Despite the significant implications of these decisions, policy was formulated and implemented within established government bureaucracy rather than through the French parliament (which had no standing committee on science and technology in the National Assembly) or other elected bodies. Further, while the Commission des Finances dealt with important topics in science and technology, it did not have staff or resources to fully study these issues. The aggressive nuclear development program was initiated and implemented by the CEA and the public utility Electricité de France (EDF).

However, the National Assembly did establish a special investigatory committee following the Three Mile Island accident. Four committee members visited the United States and published a report calling upon the French government to inform the public on contingency plans for actions to be taken in the event of a nuclear accident in France. Notably, they did not call for a full-scale assessment of safety considerations for the French nuclear program.

The methods and conclusions of the CEA and the EDF continued to be conducted out of sight of the public. The tradition of the National Assembly was to consult outside experts rather than to conduct internal reviews. In the case of the Three Mile Island report, the Assembly sought comments on the investigative report from the Académie des sciences, rather than engage in their own internal review. Throughout the formative years of nuclear energy policy this posture of the National Assembly endured through inertia from within the Assembly and absence of challenges from the academic community.

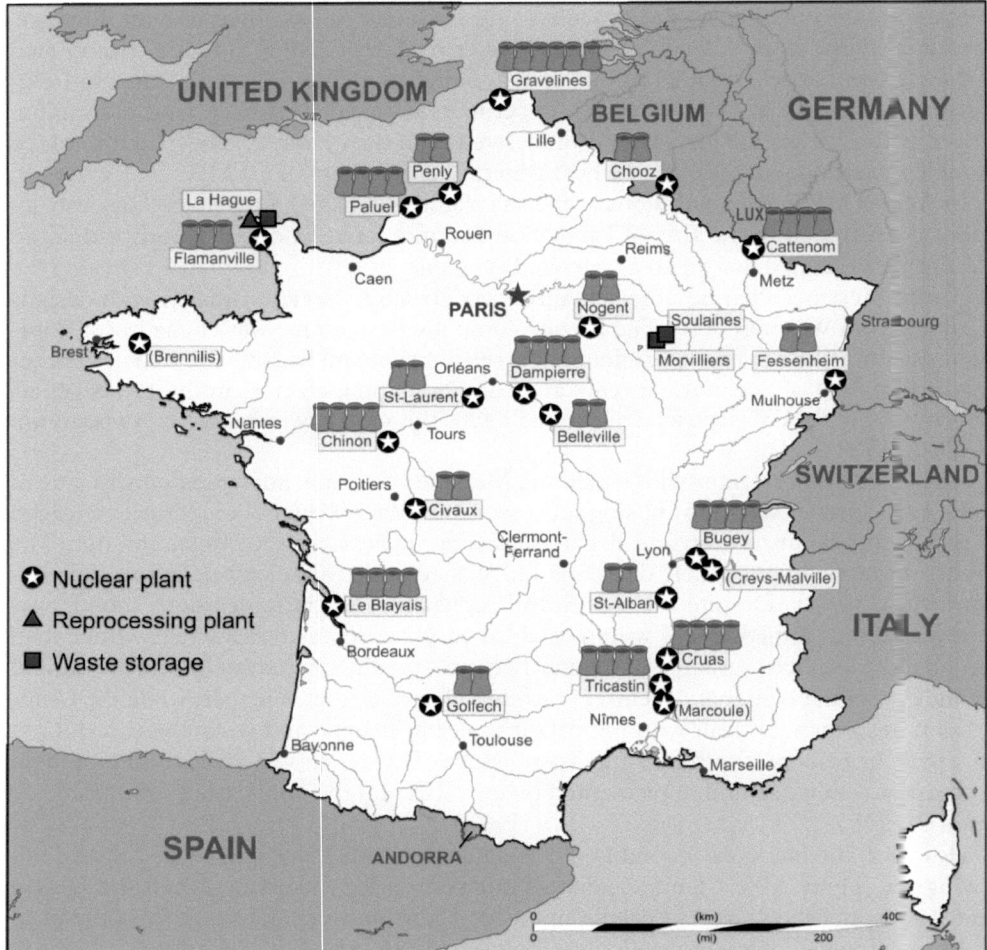

Figure 7.1 *Map showing 59 civilian nuclear power facilities in France.*

Adapted from Eric Gaba. (2008). Wikimedia Commons. https://commons.wikimedia.org/wiki/File:Nuclear_power_plants_map_France-fr.svg. under a Creative Commons Attribution-Share Alike 2.5 Generic (CC BY-SA 2.5).

The 1986 Chernobyl disaster brought into focus the role of the government in safe-guarding the health of the French people when confronting the dangers of nuclear radiation. A cloud of radioactive fallout traveled across Europe eliciting varying responses from countries in its path. Most worrisome was the response of the Soviet Union, which attempted to cover up the extent of the problem and which exacerbated the situation. The prevailing winds brought the dangerous effluent over the Netherlands, Germany, and the easternmost parts of France on its way to the United Kingdom. A highly visible response occurred in Germany, where the government distributed iodine pills and warned the public not to eat fresh produce. Judging that there was very little exposure in France, the government did little to minimize any impact or to provide guidance to the public.

A myth became widely circulated that Professor Pierre Pellerin, director of the Central Service for Protection against Ionizing Radiation (SCPRI) stated that "The Chernobyl cloud stopped at the French border." While the International Atomic Energy Agency (IAEA) later issued a report that the French response was proportional to the relatively low danger, this myth endured and helped strengthen a belief that the French government could not be trusted on issues relating to science and public policy. This myth was often cited during the pandemic as a reason that the French government could not be trusted on advice relating to Covid.

The Messmer Plan led to steady development of additional nuclear reactors through the 1970s and 1980s and reached a plateau in the 1990s, which has been maintained since 1997. With a steady level of more than 55 operating nuclear reactors, France produces approximately 70% of its electrical energy from nuclear power, which is the largest percentage of any country in the world (Figure 7.1). Stability in this supply has been a consequence of maintaining similar designs of PWRs. Despite a brief period of post-Fukushima disenchantment with nuclear energy, France continues to support a robust nuclear energy program.

Further Reading

Gabrielle Hecht, *The Radiance of France: Nuclear Power and National Identity after World War II*, MIT Press, 1998.
A history of reactor development in France during the twentieth century. This technological study is done with attention to cultural and political history.
"Nuclear Power in France," *World Nuclear Association* (last modified February 4, 2025). https://world-nuclear.org/information-library/country-profiles/countries-a-f/france.aspx
A history of nuclear power development in France.

8

Reactor Development in Russia

The living will envy the dead.

— Nikita Khrushchev[1]

Early Nuclear Research in the USSR

The Soviet Union made the study and pursuit of science a high priority. Science study and research were well supported from the early days following the Russian Revolution. Central to the pursuit of science was the USSR Academy of Sciences, established in 1925 in Leningrad before being moved to Moscow in 1934. It consisted of 250 research institutes employing thousands of full-time researchers. There were several specialized research institutes, the first being the Ioffe Physico-Technical Institute founded in 1918. In 1922 the State Radium Institute was established.

There existed a tight connection between scientific theories and communist ideology with some fields falling under the sway of theoretical dogma. This was most notorious in genetics, when in 1936 Lysenkoism embraced the notion that personality traits acquired during one lifetime could be passed onto offspring. This false construct was not renounced until Leonid Brezhnev assumed power in 1964. These unfortunate events illustrated the negation of the scientific process by ideological control that manifested itself during the Soviet era.

Soon after the discovery of fission by Otto Hahn in Germany, physicists in Russia proved that a chain reaction was possible. On December 25, 1946, the F-1, a research reactor, was activated in Moscow. It was an air-cooled, graphite-moderated reactor that used natural uranium, similar in design to the Fermi 1 reactor that had ushered in the nuclear age in Chicago in 1942.

Obninsk: The World's First Nuclear Power Plant

On June 26, 1954, in Obninsk, a nuclear reactor generating five megawatts (MW) of electrical power became the world's first reactor connected to a civilian power grid. The Soviet Union proudly discussed this achievement at the first United Nations conference

[1] Nikita Khrushchev, Speech to the UN General Assembly, September 18, 1959.

Nuclear Energy. Edward A. Friedman, Oxford University Press. © Edward A. Friedman (2025). DOI: 10.1093/9780198925811.003.0008

on the peaceful uses of atomic energy held in Geneva in the summer of 1955. This accomplishment established Soviet scientists and engineers as among the world leaders in nuclear energy. This reactor ran without difficulty until 2002.

Nestled in the "science city" bearing its name, the Obninsk reactor served as a testbed for development of future graphite-moderated water-cooled reactors and the facility became a training center for Soviet nuclear engineers. It was used for experiments and tests and was the forerunner of the graphite-moderated nuclear power reactor (RBMK) used at Chernobyl (see Chapter 12). The facility hummed along until its decommissioning in 2002, leaving an indelible impact on the Soviet nuclear industry.

The RBMK Design

The RBMK-1000's design was finalized in 1968. At the time it was the world's largest nuclear reactor design, being 20 times larger by volume than contemporary western designs. There was considerable pressure on the central planners to develop a unique Russian design and to implement expanded production of nuclear energy. The government authorized the construction of four massive RBMK-1000s before the reactor design was finalized, let alone approved. However, significant design flaws lurked beneath the surface, sowing the seeds for the Chernobyl disaster two decades later.

The RBMK design consisted of a huge cylindrical graphite block about 21 feet high and about 36 feet in diameter. Graphite serves as a moderator for the fission process. Into this cylinder there are 1,661 cylindrical channels that house fuel rods through which cooling water circulates. Of the 1,661 channels, 211 are reserved for control rods. Thus, each fuel rod is individually cooled with pressurized water that flows upward through the tube and emerges in boiling condition from the top of the tube. A crane suspended over the top of the reactor is used to extract individual fuel rods from the reactor so that fuel can be processed without interrupting the overall operation of the reactor. This system facilitates the extraction of plutonium for weapons as an ongoing process while the reactor supplies power to electricity-generating turbines.

The designers viewed the localization of fission in individually separated tubes as a safety feature. With an eye to minimizing the cost of construction, a decision was made to operate the reactor without the benefit of a containment structure. Building a containment enclosure for such a large reactor would have doubled the cost of construction.

Several flaws were recognized in the design as development proceeded but due to the early authorization to build these units and the authoritarian nature of decision making, the construction continued without the needed modifications. The most serious of the design flaws was the positive feedback that could occur if the reactor overheated, which would cause the water to boil with the introduction of air bubbles. These air bubbles replaced water that previously modified the flow and absorption of neutrons. These neutrons now instead enhanced the fission process, resulting in an increase in the heating causing the problem.

The RBMK was a unique design that was promoted by the leadership in Russia as the "Russian Reactor." The government positioned the RBMK as competition for pressurized-water reactors (PWRs) (the most common design outside of Russia), which were known as the "American Reactor."

Central planners found the RBMK an attractive design since it could be easily manufactured at a low cost and provide a large energy output. The first RBMK-1000 went into operation near Leningrad in 1973 and the second unit came online in 1975. Over the next ten years a total of 14 RBMK-1000 became operational in Russia. There was also a unit built in Lithuania that produced 1500 gigawatts of electricity.

The flaws in the design of the RBMK were kept secret and the manuals for operators were vague on important operational and maintenance issues. Also, when problems arose at a given site, the lessons learned were not shared but were also kept secret. This culture of central control and secrecy set the stage for the tragedy that befell the RBMK at Chernobyl in 1986.

Post-RBMK Designs

Maintaining a leadership position in the development of nuclear energy was a high priority for the Russian government. The highest levels of the state provided significant resources for the development of nuclear technology. In 1956 the Soviets were exploring ten different nuclear reactor designs. This approach of simultaneously exploring multiple designs became a hallmark of the Russian nuclear program.

The next two reactors built in the Soviet Union were of an unusual experimental design that was not repeated. These reactors used water at what is known as a supercritical temperature—a high temperature at which there is no distinction between the liquid phase and the gaseous phase. The motivation for this construction was the higher efficiency for heat transfer when using a system at supercritical temperature. These reactors were built in 1958 in Beloyarsk with outputs of 600 MW and 885 MW.

Nuclear-powered Icebreakers

In 1957, the USSR announced that it had completed construction of a nuclear-powered icebreaker, the *Lenin*. The *Lenin* entered into service in 1959 and provided access to Northeast Passage runs alongside the Russian Arctic coast for 4,375 miles across 11 time zones—an area that had hitherto been an isolated region of the Soviet Union. Traffic through this Arctic passage had previously been limited to seasonal travel.

The *Lenin* went into operation as the world's first civilian nuclear vessel. It was joined in ice breaking activity in 1975 by a second ship, the *Arktika*, and then a third in 1977, the *Sibir*. While the first-generation icebreaker, which became operational in 1958, was a PWR using 5% enriched uranium, the second-generation reactors that became operational in 1970 used 90% enriched uranium. The latter was thus employing weapons-grade materials. The success of these vessels led the Soviets to build a fleet of icebreakers that opened this region to economic development.

Following the October Revolution, the Soviets had undertaken a large-scale study of the Arctic Ocean. They built a network of research stations and sent expeditions into the area. Through the years they established a far-flung network of settlements made viable by the development of their nuclear-powered icebreakers.

VVERs

PWRs were being developed for icebreakers alongside a major initiative to develop PWRs for urban locations. In 1957 in the city of Novovoronezh construction began for a PWR that became the first in a series of reactors known as VVERs ("water–water energetics"). This first VVER became operational in 1964 with output power of 210 MW electric. This reactor, for which water served as both coolant and moderator, became the model for a second track of Russian reactors joining the emerging fleet of RBMKs being used for civilian power production. The Soviet planners promoted these two designs as competitors in the world of Russian nuclear energy production.

The VVER-210 was followed with the VVER-365 and the VVER-440 (with numbers that noted the electrical output), none of which were protected with containment structures. A larger VVER-1000 was built in 1975 that incorporated a containment structure.

The VVER design became a mainstay for the export market with more than 30 reactors successfully installed in countries in Eastern Europe and Asia. The VVER is also widely used within Russia. Production of the VVER has been constrained by the limited number of facilities in Russia that have the capacity to fabricate the metal enclosure for the reactor core. Manufacturing considerations also need to consider limitations of train tunnel size through which the reactor core enclosure must pass when shipped between the fabrication center and the assembly location. Also, the final product needs to be sized for transport through train tunnels while enroute to export locations.

The Russian drive to be successful in exporting nuclear technology created a need to adhere to international design standards for safety. This requirement was met by Russia with respect to VVER designs when units were sold to Finland for use in the town of Loviisa in 1977 and 1980.

The safety record for VVER reactors both in Russia and in other countries has been excellent.

Challenges of Central Planning

Development of nuclear reactor technology in the Soviet Union proceeded in general society as it did with ice breakers. All major industrial enterprises in the Soviet Union took place as coordinated initiatives of central planning. After the construction of three nuclear reactors in the 1960s there was a massive initiative promoting nuclear energy in the 1970s with the construction of 12 reactors, followed in the 1980s with the introduction of an additional 21 units. There would have been more had the Chernobyl disaster not occurred in 1986.

Economics and development in the Soviet Union proceeded through a series of five-year plans. Although in principle central planning allows for complete coordination of the economic system, it is a challenge for planners to successfully oversee all aspects of a complex development program.

Despite the impressive expansion of nuclear energy in the Soviet Union during the final decades of the twentieth century, the achievement fell far short of the planned development. Between 1970 and 1975 only 60% of the planned additional energy units were completed. In 1981, three 1-million MW reactors failed to be completed on schedule.

An advantage of a central planning structured economy is that it allows complex construction and fabrication initiatives to proceed in a coordinated fashion. Most notable for the nuclear development sector of the economy was the establishment of Atommash in 1976 as a multidisciplinary engineering company. The city of Volgodonsk was chosen for this enterprise. It is in the south of Russia with access to river and lake waterways that allow the easy shipment of heavy equipment. Roads, railways, and an airport were built to support this enterprise. A nuclear power plant was built to supplement the available energy needed for production.

The Legacy of Atommash

The Chernobyl disaster of 1986 and the dissolution of the Soviet Union in 1990 decreased the number of orders for nuclear equipment. The company had to expand its assortment of goods to continue functioning. This change in organization was not successful and in 2000 Atommash was restructured and began manufacturing a wide array of technical equipment.

Included in its portfolio were gas turbines, equipment for the Russian space program, and some renewed support for the nuclear energy sector of the economy. One of the early successes of the restructured Atommash was the manufacture of the structure needed to house a fusion reactor. Also included in their manufacturing products have been energy systems for wind power and launch pads for missiles and spacecraft.

Russia's Nuclear Industry Today

Today Russia is one of the world's leaders in nuclear energy generation. It maintains 36 operating nuclear plants that supply 20% of its electrical energy needs and includes eight REMKs, 22 VVERs, two sodium-cooled reactors, and two floating PWRs. In addition, Russia operates two low-power output RBMK designs that produce only 12 MW electric.

Russia's impressive export market has installed VVER reactors in Bangladesh, Belarus, China, Egypt, Finland, India, Iran, Türkiye, Ukraine, and Vietnam.

Rosatom is the organizational entity that oversees and manages the Russian nuclear establishment. Not only does it operate a vast installed base of nuclear power plants in Russia, but also it has the most widespread network of exported nuclear generation plants. Its dominance in the production and management of nuclear fuel has proven to be a source of international contention regarding the war in Ukraine. Nuclear initiatives in the United States, Canada, Europe, and South Korea are moving forward to challenge Rosatom's position in the world.

Further Reading

"Nuclear Power in Russia," *World Nuclear Association* (last modified December 3, 2024), https://world-nuclear.org/information-library/country-profiles/countries-o-s/russia-nuclear-power.aspx

A comprehensive summary of nuclear reactor development in Russia.

Paul R. Josephson, *Red Atom: Russia's Nuclear Power Program from Stalin to Today*, W. H. Freeman and Company, 2000.

This is the work of a Russian scholar who addresses the relationship between Soviet politics and science policy.

Sonja D. Schmid, *Producing Power: The Pre-Chernobyl History of the Soviet Nuclear Industry*, Penguin Random House, 2015.

A scholarly account of the history of the Soviet nuclear industry that is based upon interviews with key individuals and extensive research in Soviet archives.

9

Reactor Development in China

China is fully capable and obligated to promote the certification of the green attribute of nuclear energy. By taking the lead in doing so, China can enhance its voice in global energy governance.

— Lu Tiezhong, Chairman, China National Nuclear Power (CNNP),
March 18, 2024

Nuclear Development in China

This chapter traces the arc of China's nuclear development, from its early roots amid a civil war and Cold War tensions, through a period of domestic power expansion and international collaboration to the present day. It is a story intertwined with China's political evolution and its changing role on the global stage.

At the time that fission was discovered, there were two factions competing for control of China. The civil war between the Nationalist Party and the Communist Party continued until late 1949 when Chiang Kai-shek withdrew with his Kuomintang followers to the island of Taiwan with Mao Zedong and the Communist Party taking control of the mainland.

In mid-1950s China engaged in initial meetings to assess and to coordinate scientific manpower. Chinese scientists located on the mainland were organized within a newly structured Academy of Science and contact was made with Chinese scientists located abroad to return. These communications were a combination of persuasion and implied threats. At the same time promising students were identified for studying abroad.

The first Five-Year Plan covering the years 1953–1957 identified priority areas of the economy that required support from science and technology. The peaceful production of atomic energy was one of 11 fields that were singled out.

Outreach to the Soviet Union was established leading to an agreement that was formally adopted on October 12, 1954, for cooperation between the Soviet Union and the People's Republic of China (PRC) in scientific and technical areas.

In April 1955, an agreement was concluded in Moscow for the construction in China of an atomic reactor with 7–10-megawatts (MW) capacity. Also agreed to was construction of cyclotrons for use in atomic physics studies. The first nuclear reactor to become

Nuclear Energy. Edward A. Friedman, Oxford University Press. © Edward A. Friedman (2025). DOI: 10.1093/9780198925811.003.0009

operational in China was a research reactor that used heavy water as a moderator and natural uranium as fuel. This facility, built in a suburb of Beijing, first became critical in June of 1958. This was a major contribution from the Soviet Union.

Breakup with the Soviet Union

In 1957, Soviet leader Nikita Khrushchev announced that Soviet support to China would be extended to include weapons production. However, in 1960 Soviet experts were withdrawn from China and Soviet support to China for both military and civilian applications of nuclear energy were terminated.

The breakup between the Soviet Union and the PRC evolved in the early 1960s due to several factors. The ideological divide between the two communist powers played a significant role. The Soviet Union, under Khrushchev's leadership, pursued a policy of de-Stalinization and advocated for peaceful coexistence with the capitalist West. In contrast, Mao's China adhered to a more militant and revolutionary form of communism. Mao criticized Khrushchev's approach as revisionist and accused the Soviet leadership of betraying true Marxist–Leninist principles. This difference played into the competition between the two countries for leadership in the communist world, with China wishing to establish itself as the center for world revolution. China's leaders were also concerned about the Soviet influence in Mongolia, Xinjiang, and other border regions that were traditionally within the Russian sphere of influence. These and related differences led to a complete break in relations.

China's Pursuit of the Bomb

Following the explosion of an atomic bomb by China in 1964 US intelligence initially assumed that the bomb used plutonium produced at a reactor near Paotow in inner Mongolia. Given that China was prohibited by their agreement with the Soviet Union from using the reactor located near Beijing for weapons production, it was also assumed that the plutonium for their bomb came from the secret Paotow reactor. That reactor was discovered through aerial reconnaissance conducted by the United States.

However, analysis of the fallout from the bomb revealed that the design was modeled after the Hiroshima weapon that used U-235 as the fissionable material, rather than the Nagasaki weapon that had used plutonium. The conclusion was that the fuel for the Chinese explosion originated at the Lanzhou Nuclear Fuel Complex, where China had built a gaseous diffusion plant. Prior to the explosion US intelligence had not realized that the plant was operational. Given the limited capacity of the Lanzhou plant it was further assumed that China had used a two-step approach to uranium enrichment, with diffusion as step one and electromagnetic separation as step two, as had been done at Los Alamos in 1945.

The Secret 816 Project

In the 1950s, when relations with the Soviet Union were deteriorating, China decided to safeguard its capability of producing plutonium from uranium that had undergone fission in a nuclear reactor. A formidable reactor construction project was initiated in secret in Fuling, about 1100 miles west of Shanghai. The decision was made to construct the reactor underground in Jinzi Mountain. Premier Zhou Enlai personally approved this plan in 1966. A design was implemented that located a reactor that was an exact replica of an existing water-cooled reactor, known as the 404 Project, that had been provided to China by the Soviet Union. The chosen location was deep within the mountain surrounded by a heavy rock formation that could withstand a major bomb attack and a level-8 earthquake.

A workforce of 60,000 people was enlisted in construction of what became the world's largest artificial cave. It was a top-secret project with many working at the site unaware of its true purpose. The complex included schools, a market, and a hospital so that workers could live in total isolation. The nearby town of Baitao disappeared from the map.

The evolution of China's nuclear energy program and changes in international relations led to the project being abandoned at a point in 1984 when it was 85% complete. Remarkably, in 2010, this huge construction project was opened to the public as a tourist attraction! The history of what was known as the 816 Project was finally revealed. Tourists were escorted into a football field-sized cavern intended to house the reactor, which now had neon lights intended to illustrate the planned location of uranium fuel rods. It stands as a striking relic of a bygone era in China's nuclear journey.

Transition to Domestic Nuclear Power

China's first nuclear reactors were developed with the objective of producing plutonium for weapons. Beginning in the 1950s, Mao gave China's military establishment the green light to develop nuclear weapons and ensure, in his view, that China would not be blackmailed by nuclear armed imperialist adversaries. China was by far the poorest and least developed of the states that developed nuclear arms in the years after the Second World War. The military defeated advocates of nuclear power in bureaucratic battles waged over the allocation of China's limited resources for nuclear research and development.

Pursuit of commercial reactors for domestic power began in the 1980s. The initiation of a nuclear plant for domestic power originated in Hong Kong with British businessman Lawrence Kadoorie, who owned CLP Power. He proposed a plant to be based in nearby Guangdong province that would provide power to both Hong Kong and an area of southern China. Then British Prime Minister Margaret Thatcher was enthusiastic about the proposal and committed support for its implementation from the UK Department of Industry. Chinese leader Deng Xiaoping was also enthusiastic about the plan. In 1985, Deng and Kadoorie met to endorse the plan and initiate construction.

In Hong Kong the proposal met with strong opposition precipitated by the Chernobyl catastrophe on April 26, 1986. Over time, more than a million people (one fifth of Hong Kong's population) signed a petition opposing nuclear power. Over a hundred community groups initiated discussions on this construction with an emphasis on environmental issues. This collaboration was also seen as a threat to independence.

The concerns of Hong Kong's citizens were treated in a high-handed manner. China's minister of nuclear industry Jiang Xinxiong rejected the petition with the response that "The Chinese government will treat the matter with a scientific and practical attitude. It will not halt the Daya Bay project because of the objections from some people."

However, the Chinese administration did respond to citizens' concerns by following up with care on construction issues that arose. They also emphasized that, unlike Chernobyl, there was a sealed containment building surrounding the reactor. Great care was exercised in training the staff for the reactor and 100 Chinese engineers went to France for training and the acquisition of a control room simulator from the United States for use in local training programs.

This project went forward through a contract with Framatome, the French company that designed and built the reactors. Two identical pressurized-water reactors (PWRs) were built, each with an output of 944 MW electric. The first unit began operations on August 31, 1993, and the second unit on February 2, 1994. About 70% of the output from the Daya Nuclear Power Plant was used in Hong Kong, with the remaining 30% being directed to Guangdong on the mainland.

Growth of China's Nuclear Sector

The first operational indigenous nuclear power reactor was the CNP-300—a PWR with a power output of 300 MW electric. This was not only the first domestic design but also the first Chinese reactor to be exported. Installation of a CNP-300 became operational at Pakistan's Chashma Nuclear Power Plant in 2000. Its implementation in China was in the town of Qinshan, located about 78 miles southwest of Shanghai. Construction began on March 20, 1985, with the reactor becoming operational on April 1, 1994.

The Daya Nuclear Power Plant was constructed two years later in 1987 with the two units coming online in 1993 and 1994. There was a hiatus of several years in reactor construction before Qinshan 2 began in mid-1996, more than a decade after the construction started for Qinshan 1. With an output of 610 MW electric, Qinshan 2 had twice the power capacity of Qinshan 1. This began a surge in nuclear power plant construction that saw eight plants built beginning with Qinshan 2 in 1996 and continuing until Ling Ao 2, for which construction began in 1997.

There then occurred a second hiatus that continued for six years before an intense period of power reactor construction began in 2005 and extended until 2012, during which time ground was broken for the building of 29 nuclear reactors with capacities in the 1,000-MW range. The last reactors built during this surge were constructed at Fuqing on the southeast coast of China about 110 miles across the strait from Taiwan.

These were the last two "Generation II" reactors built in China. Their construction started in 2009 and 2011.

Most of the reactors built in China prior to 2010 were PWRs modeled after the French design. There were also two heavy-water-cooled Canada Deuterium Uranium (CANDU) reactors that originated in Canada. These were installed as extensions to the Qinshan power station: Qinshan 3 went online in 2002 and Qinshan 4 went online in 2003, each with a power output of 677 MW.

International Collaborations and Tensions

China sought state-of-the-art nuclear technology from countries in the West. In addition to France and Canada, a nuclear information treaty was formalized with the United States in 1985. This arrangement was the first such agreement for the United States with a communist country and the third established with a nuclear weapons state. The United States had previously entered into agreements with Britain and France through a treaty established with the European Union.

Given the competitive position of China vis-a-vis the United States on the world stage, a treaty between the two countries on sharing of nuclear technology was surprising. In a letter of transmittal for this agreement, President Ronald Reagan wrote: "Based on our talks with the Chinese, we can expect that China's policy of not assisting a non-nuclear weapon state to acquire nuclear explosives will be implemented." An unusual level of trust was implicit in this agreement. Reagan also noted that "China has said that it will require IAEA safeguards on its future nuclear export commitments to non-nuclear weapons states."

This agreement led to subsequent relations and protocols, but it was not until May 2010 that a Memorandum of Further Cooperation was signed between the US Nuclear Regulatory Commission and the PRC that authorized construction of a Generation III reactor designed by Westinghouse to be built in China.

In addition to collaborations that support the development of new technology, relationships that foster intellectual development of individuals are of great value. Through the years many Chinese students and scholars have studied in the United States. In 1980 there were 80,720 students from China studying in the United States. This number rose to 439,702 in 2015 and has since declined to 289,526 in 2023.

In addition to Chinese citizens studying in the United States through individual initiative, there have been many institutionally organized exchange programs. In the area of nuclear learning, research and development has been a program between The Nuclear Power Institute of Texas A&M and Harbin Engineering Institute in China. That program sponsored student and faculty visits between the collaborating campuses and joint study programs. With the cooling of relations between China and the United States, the number of Chinese citizens studying in the United States has been declining and programs like that between Texas A&M and Harbin have been canceled.

In November of 2018, the US Department of Justice launched a program to identify and prosecute Chinese spies within the academic community or those pursuing economic espionage. This was a flawed initiative that failed to identify relevant wrong-doing, and which resulted in activating misguided prosecutions and in promoting racial profiling that engendered fear and mistrust within the scholarly community. Thousands of investigations were pursued and of the 50 indictments, a handful resulted in convictions for misstatements on legal filings and tax returns for personal gain, rather than espionage.

While the program ended in February 2022, its impact has engendered an atmosphere of suspicion that has continuing impact. It has resulted in students and scholars from China often choosing to pursue their research interests in countries other than the United States.

Note that this ambivalence toward China, and the suspicion of Chinese scholars as communist agents, dates to early days of China's communist regime. In 1950, a leading scholar, Qian Xuesen, who had received advanced degrees in aeronautics from MIT and CalTech in the 1930s, and became a full professor at CalTech in 1949, was accused by the US Government of being a communist agent. Utterly without foundation, this accusation arose during a period of red scare hysteria in the United States, and Qian was stripped of his security clearance and he and his family were subjected to partial house arrest and surveillance for five years. In 1955 he was released and exchanged for American pilots who had been captured by China during the Korean War. Upon his return to China, he became a leader of the Chinese space program and a key member of the scientific community.

Unfortunately, the sentiments that gave rise to the Qian debacle continue to influence US–China scientific and technological relations more than 75 years later. This tension has spillover effects in related areas and continues to complicate efforts at collaboration and information sharing.

Looking Ahead

China's nuclear journey has been marked by starts and stops, by grand ambitions and pragmatic shifts, and by the constant interplay of domestic priorities and global dynamics. As China looks to the future, with plans for continued expansion of its nuclear power sector and a growing role in global nuclear leadership, the lessons and legacies of its nuclear past will continue to shape its path moving forward.

Further Reading

"Nuclear Power in China," *World Nuclear Association* (last modified January 14, 2025) https://world-nuclear.org/information-library/country-profiles/countries-a-f/china-nuclear-power.aspx
A review of nuclear power development in China.
Xu Yi-Chong, *The Politics of Nuclear Energy in China*, Springer, 2010.
The author is a scholar at the Griffith University in Australia and explores nuclear development in China in the context of the political environment.

10
Reactor Development in Other Countries

I declare the transition to new energy sources and the start of a new 1000 year history here in Saemangeum (South Korea) today.

—Moon Jae-in October 30 2018[1]

Overview

In 2023 there were 412 nuclear reactors providing electric power in the world. Countries running 251 of these reactors were discussed in prior chapters (93 in the United States, 56 in France, 55 in China, 38 in Russia, and 9 in the UK). This chapter presents an overview of the nuclear reactor story for Canada (19 reactors), Japan (33 reactors), South Korea (25 reactors), and India (22 reactors). Together with these 99 reactors, there are an additional 62 reactors distributed among 22 other countries around the world.

In 2023 fewer than a dozen countries had reactors under construction, with the largest number in China (with 21). Chapter 24 discusses in detail the world export market for nuclear reactors that was dominated in 2023 by Russia and China, with the United States, South Korea, and France becoming increasingly competitive.

Canada

Canada had close relations with both Great Britain and the United States at the time of the development of atomic weapons in the 1940s. Given the vulnerability of locations in Britain to aerial attack from Germany, Canada became a preferred location for development of laboratory-based activities that supported weapons development. One of these

---

[1] President Moon Jae-in, Remarks by President Moon Jae-in at Saemangeum Renewable Energy Vision Declaration Ceremony, October 30, 2018, https://www.korea.net/Government/Briefing-Room/Presidential-Speeches/view?articleId=164959.

Nuclear Energy. Edward A. Friedman, Oxford University Press. © Edward A. Friedman (2025). DOI: 10.1093/9780198925811.003.0010

was heavy water as a moderator that would facilitate a chain reaction of natural uranium in a nuclear reactor, thus supporting plutonium production.

The first nuclear reactors that became operational were constructed in the United States using graphite moderators with natural uranium, while the first reactors to use heavy water were constructed in Canada.

Ordinary water contains two hydrogen atoms and one oxygen atom, while heavy water consists of deuterium atoms instead of hydrogen atoms. In deuterium, the nucleus has a neutron and a proton. Heavy water has the capacity of facilitating a chain reaction because deuterium does not absorb neutrons, as is the case for water because the deuterium molecule already contains a neutron.

Heavy water occurs as a small percentage of ordinary water and is extracted in plants that take advantage of its slight difference in boiling point from that of ordinary water. Also, the compound hydrogen sulfide exhibits slightly different chemical properties from that of deuterium sulfide with which it occurs naturally. This difference can also be exploited to achieve separation.

In 1944 a decision was made to build a heavy-water moderated reactor in Canada, which resulted the Zero Energy Experimental Pile (ZEEP) reactor, which became operational at Chalk River in 1945. This 10-watt reactor was the first self-sustaining chain reaction outside of the United States. The ZEEP reactor was used for basic research until 1970. It was dismantled in 1997 and is commemorated in the Science and Technology Museum in Ottawa, Canada.

The ZEEP reactor led the way for development of other heavy-water reactors. This evolution of designs led to a successful line of reactors known as Canada Deuterium Uranium (CANDU) reactors. An attractive feature of the CANDU reactor is found in the fuel rod design. In a CANDU reactor the fuel is bundled into a collection of tubes that allows refueling to take place in individual tubes without shutting down the reactor (as is the case with most pressurized-water reactors, PRWs). This allows the CANDU reactor to be used more efficiently than most other reactors.

Canada has 19 CANDU model reactors in operation with power output ranging from 516–881 megawatts (MW) electric. There have also been six reactors shut down that had CANDU-related designs.

Canada had an aggressive program that promoted the export of CANDU reactors in the 1970s–1990s. Sales were successful in China (2), Argentina, South Korea (3), Pakistan, Romania, and India. Further, it was revealed that in fabricating its first nuclear weapon in 1974, India used plutonium that was provided by the CANDU reactor from Canada. Canada has been the recipient of much criticism for selling reactors to dictators and unreliable countries.

India

India has an ambitious nuclear development program with an aim of being self-sufficient. Following the acquisition of two small CANDU pressurized heavy-water reactors from Canada in 1972 and 1980, Canada initiated an indigenous pressurized heavy-water reactor development program that led to the construction of 19 reactors of that design. India also obtained two PRWs from Russia and two boiling-water reactors (BWRs) from the United States, which were built by General Electric.

India has a long-term goal of building thorium-based reactors. This quest is driven by a desire to exploit the advantages of a thorium reactor design and the fact that India has the world's largest deposits of thorium, while having quite limited deposits of uranium.

Thorium cannot sustain a chain reaction itself, but the most common isotope of thorium, which is 232, transforms into uranium-233 (U-233) when it absorbs a neutron, which does participate in an energy-producing chain reaction. A reactor based on thorium does not produce long-lived radioactive waste since it does not contain uranium-238 (U-238) or the heavy radioactive isotopes produced by neutron absorption in U-238

India has not yet succeeded in constructing a thorium-based reactor, which requires a source of neutrons with which to start the process. The challenge of constructing a thorium-based reactor continues to be pursued by India, the United States, and elsewhere.

India's detonation of a nuclear bomb in 1974 triggered the establishment of a multinational nuclear materials export control group, and 48 countries joined for this purpose. The alliance was named the Nuclear Suppliers Group (NSG). Membership requires that a nation be a signatory to the Nuclear Non-Proliferation Treaty.

To fully support its civilian nuclear energy program India has been eager to eliminate the constraint of non-NSG membership on nuclear-related trade. In 2008, through diplomatic initiatives of the United States, India was granted a partial trade waiver from NSG controls but more recently has been denied full NSG membership. This denial is based on the unwillingness of India to sign the Non-Proliferation Treaty with an obligation to give up its nuclear arsenal.

In August 2024, India's National Department of Atomic Energy and Tata Consulting engineers announced plans to redesign their existing 220 MWe pressurized heavy-water reactor to develop the Bharat Small Modular Reactor. They plan to take advantage of three-dimensional design platforms that were not available when these reactors were first fabricated 40 years ago. They expect to achieve a high degree of standardization that would allow modular, scalable, safety-aligned production with the goal of making 40–50 such reactors in fewer than seven or eight years. Their goal is to facilitate India's achieving net-zero carbon emissions by 2070 or earlier. This action illustrates the awareness of major nations that nuclear energy will be a necessary component in the quest to achieve a carbon neutral world.

South Korea

Nuclear research and development cooperation between the United States and South Korea dates to the beginning of President Dwight D. Eisenhower's Atoms for Peace program. The first major US—South Korea nuclear project, a 100-kilowatt (KW) reactor, began operating in 1962 and was later upgraded to 250 KW and finally to 2 MW.

These joint activities were carried out under a series of peaceful cooperation agreements signed between 1956 and 1965.

The war between North and South Korea was fought with active engagements between 1950 and 1953. The fighting ended with an armistice on July 27, 1953. The two sides never achieved a peace accord. Given the ongoing tension between the two sides, it is unsurprising that, in the mid-1970s, the South Korean dictator General Park Chung-hee initiated a secret push to develop nuclear weapons before the end of the decade. In 1974, US Secretary of State Henry Kissinger learned that South Korea was negotiating with France to purchase a chemical separation plant that could be used to produce plutonium from spent nuclear fuel, which could be used to construct a bomb.

In a meeting in October 1975 between the United States and South Korea, America raised strong objections to the South Korean negotiations with France. The main point was that if the reprocessing plant were to operate for 12 months, it could produce 20 kilograms of plutonium which would be enough to fuel three "Fat Man" sized nuclear bombs, like that used to destroy Nagasaki. Given these objections, South Korea canceled negotiations with France for a reprocessing plant.

At the same time, South Korea had embarked on installing its first nuclear power plant, Kori 1, which was built under contract with Westinghouse. Construction of this 576-MW electric reactor started in Busan in 1971 and began commercial operations in 1978.

South Korea continued with a robust program of nuclear reactor development. In the 1970s and 1980s South Korea started construction on six PRWs purchased from Westinghouse and two from France. This followed in the 1990s with three CANDU reactors from Canada and seven additional PRWs that were built by US contractors with gradually increased involvement of the indigenous Korea Power Engineering Company (KOPEC).

As South Korea developed a nuclear power export capability it felt increasingly constrained by the agreement that was originally developed in 1974, which required US consent to engage in any enrichment activities. The agreement between the two countries that controls the availability and processing of enriched uranium was revised in 2015.

Under the terms of the 2015 agreement the United States agreed to supply South Korea with enriched uranium it needs for its nuclear reactors. The agreement leaves open the possibility that South Korea might be allowed to enrich uranium to less-than weapons-grade levels, whereas such enrichment had been strictly forbidden in the 1974 agreement.

The new agreement takes into consideration that the South Korea nuclear industry has developed to an advanced stage as an exporter of nuclear technology.

Japan

As the only country to have experienced a nuclear weapons attack, there was an initial negative attitude in Japan regarding the use of nuclear energy. This anti-nuclear-energy disposition was reinforced in 2011 by the Fukushima accident. Yet the pressing need in Japan for carbon-free energy is supporting a return to robust nuclear energy policies.

Following the initiation of the Atoms for Peace program in 1953, the United States began encouraging the use of atomic energy in Japan. In 1954 government infrastructure was established in Japan to promote nuclear initiatives. The first nuclear power plant was commissioned in Japan in 1966. This was a Magnox design from the United Kingdom that ran on natural uranium, with a graphite moderator and carbon dioxide coolant. This reactor had an output of 150 MW electric.

The second reactor built in Japan was a BWR designed by General Electric and built in collaboration with Hitachi. This reactor at Tokai began construction in 1973 and went online in 1978 generating an output of 1060 MW electric.

Up until the Fukushima accident in 2011 (see Chapter 13), Japan generated approximately 30% of its electrical energy from nuclear reactors. More than 30 reactors were operational with a mix of BWRs built by GE–Hitachi and PWRs built by Mitsubishi Heavy Industries. Mitsubishi used designs obtained from Westinghouse.

In addition to the Fukushima accident, there were other earlier serious accidents. For example, in December of 1995 a fire broke out at the sodium-cooled Monju Nuclear Power Plant. A pipe carrying sodium coolant ruptured and exposed the sodium to the air and moisture in the facility. This caused a fire that produced intense heat that warped several steel structures in the room. A full shutdown was postponed for 90 minutes, during which time significant damage occurred. There was considerable public outrage when it was revealed that the operating company had tried to cover up the extent of the damage. The cover-up included falsifying reports and editing a videotape taken immediately after the accident. Also, a gag order was issued to stop employees from revealing that the tapes had been edited.

The official in charge of the investigation committed suicide by jumping from the roof of a Tokyo hotel. The scandal surrounding this event delayed the plant's restart until 2010. When it was seen that more investment was needed to enable the reactor to continue for another ten years, a decision was made to close the facility in December of 2016.

In September 1999 at the Tokaimura Nuclear Plant, mishandled nuclear fuel went critical and exposed 667 people to heavy doses of radiation that resulted in the death of two individuals. It was determined that the accident was due to inadequate regulatory oversight, lack of appropriate safety culture, and inadequate worker training

In August 2004, an accident in a building housing the turbines for the Mihama Nuclear Power Plant suffered leaking of hot water and steam from a broken pipe with the result that five people died and others were injured. While this was a serious accident, it did not involve the reactor core.

Other Countries

In addition to the nine nuclear reactor manufacturing countries discussed thus far there are at least seven that have either built reactors or collaborated in the building of a reactor. These include Spain, Finland, Czech Republic, Sweden, Argentina, Belgium, and Brazil.

Note that Siemens, which had been a manufacturer of reactors for electrical power, abandoned this line of work at the same time that the government of Germany closed its nuclear power plants.

The dynamics of the nuclear reactor export market is discussed in Chapter 24. Up to 2023 the dominant player in exporting nuclear reactors has been the Russian Federation, with the single largest recipient of its reactors having been Ukraine, which has acquired 15 reactors from Russia. The location of the reactors in the middle of a war zone has been the cause of much concern and anxiety.

Further Reading

Andrew Leatherbarrow, *Melting Sun: The History of Nuclear Power in Japan and the Disaster at Fukushima Daiichi*, self-published, 2022.

Mohd Lateef Mir, *India's Policy on Nuclear Energy Issues and Challenges*, self-published, 2024.

"Nuclear Power in Canada", *World Nuclear Association* (last modified January 7, 2025). https://world-nuclear.org/information-library/country-profiles/countries-a-f/canada-nuclear-power.aspx

"Nuclear Power in India," *World Nuclear Association* (last modified February 19, 2025). https://world-nuclear.org/information-library/country-profiles/countries-g-n/india.aspx

"Nuclear Power in Japan," *World Nuclear Association* (last modified February 4, 2025). https://world-nuclear.org/information-library/country-profiles/countries-g-n/japan-nuclear-power.aspx

"Nuclear Power in South Korea," *World Nuclear Association* (last modified May 3, 2024). https://world-nuclear.org/information-library/country-profiles/countries-o-s/south-korea.aspx

Each of these four reports by the World Nuclear Association provides comprehensive summaries of nuclear energy developments in the respective countries.

11

The Three Mile Island Accident

I once absent-mindedly ordered Three Mile Island dressing in a restaurant and, with great presence of mind, they brought Thousand Island Dressing and a bottle of chili sauce.

— Terry Pratchett

Outlook for Nuclear Energy in the 1970s

The 1970s were a pivotal decade in the boom and bust of nuclear energy in the United States. In the early part of the decade, nuclear construction was in full swing, having begun in the 1950s and accelerated through the 1960s. Optimism among the public, federal government, and utility companies for the technology's promise of abundant, pollution-free energy, coupled with rapidly growing electricity demand, drove announcements for dozens of new nuclear reactors each year. In 1973, the Atomic Energy Commission (AEC) forecasted the need for around 1,200,000 megawatts (MW) of nuclear power (roughly 1000 reactors) by the year 2000. Two years later, President Gerald Ford in his State of the Union address announced the more moderate but still ambitious goal of having "200 major nuclear power plants" operating within ten years. However, the United States achieved just half of that number.

The oil crisis of 1973 kicked off major turmoil in the utility sector, upending plans for new nuclear construction. Of the nearly 200 nuclear plants planned, by mid-decade over 120 had been delayed or canceled. Severe inflation and rising labor and material costs made it difficult for utilities to raise money for basic electricity system upkeep, let alone new nuclear projects, which were becoming more expensive with shifting regulations from the Nuclear Regulatory Commission (NRC). In its January 1975 issue, *Nucleonics Week* described the nuclear industry as being in a state of "utter chaos" and that utilities had "no idea how to finance nuclear plants."

The nuclear sector was not only fighting economic battles but deliberate and effective opposition as well. The anti-nuclear movement, which had originated with the use of nuclear weapons at Hiroshima and Nagasaki at the end of the Second World War, expanded in response to programs of weapons testing and a nuclear arms race.

Prior to signing the Nuclear Test Ban Treaty in 1963, the United States and Russia each conducted over 200 atmospheric nuclear denotations and a far greater number

Nuclear Energy. Edward A. Friedman, Oxford University Press. © Edward A. Friedman (2025). DOI: 10.1093/9780198925811.003.0011

of underground tests. These ignited widespread and long-lasting fears of radiation. One highly publicized study, with preliminary findings issued in 1961, reported finding traces of strontium-90, a radioactive isotope from the bomb testing, in the teeth of babies born in the 1950s. There remain long-lasting observable levels of this radioactive isotope in the environment. For example, in some scientific applications requiring ultra-low radiation environments, researchers must source materials, such as lead radiation shielding, mined before the nuclear weapons testing, as the persistent radioactive decay of the fission products can disrupt sensitive readings.

As reactor construction for electricity use expanded, many of the fears about nuclear weapons were grafted onto commercial nuclear energy technology, which experienced mounting opposition. Over the 1960s and 1970s, the anti-nuclear movement made significant legal and cultural headway. Litigious groups like the Natural Resource Defense Council prompted sweeping regulation on nuclear plant construction and operation. Their victories ranged from pressuring utility companies to build massive cooling towers in order to keep warm water discharge out of rivers (thereby creating imposing silhouettes on the horizon whose steam emissions are easily confused with smoke or radioactive plumes); to greatly expanding the scope of environmental reviews for new nuclear construction; to embarrassing the AEC in highly publicized hearings (such as those regarding the semi-scale tests of emergency coolant circulation systems in early 1970s that revealed major potential shortcomings in this critical safety apparatus). Civil disobedience or political organizing by concerned local groups (often with legal assistance from larger environmental groups) had also delayed or canceled nuclear construction projects in California and New York.

In the broader discourse, books such as *Perils of the Peaceful Atom* (1969) stoked fears among the public and politicians of nuclear power reactors as being "saturated with unknowns," emphasizing that "under certain circumstances . . . radioactive contents . . . a million to a billion times more toxic than any known industrial agent . . could fall out over a large area." Among its rhetorical devices were long lists of monstrous-sounding genetic mutations that could be caused by excessive radiation exposure. Climate change was not a major concern at the time, but the authors' prescription for meeting the country's growing energy needs involved burning its "near limitless" coal reserves for as long as needed to find a renewable solution. As the anti-nuclear movement matured, its arguments against nuclear power became more sophisticated and persuasive, merging radiation and safety concerns with structural critiques of the utility sector. Disillusionment with large, centralized energy and industrial systems gained ground with books like E. F. Schumacher's *Small Is Beautiful* (1973) and Amory Lovins's 1976 article "Energy Strategy: The Road Not Taken" in *Foreign Affairs* that outlined a vision for a more secure and benign "soft energy path" incompatible with large utility generators.

Proponents of nuclear power defended the technology against these claims. However, having enjoyed substantial public approval during the 1950s and much of the 1960s, nuclear energy was quickly losing support in the public arena. The combatting forces for and against nuclear energy created a roller coaster for the sector: over 100,000 MW of new reactors were announced in the first half of the 1970s, only to be canceled over the remainder of the decade and the first half of the 1980s. Before the end of the decade, the decline of new nuclear construction was precipitous, a trend only reinforced by an event that gripped the nation in fear, sent the anti-nuclear movement soaring, and delivered a fatal blow to aspirations for new nuclear energy for the rest of the twentieth century: the nuclear accident at Three Mile Island, Pennsylvania.

The incident at the Three Mile Island nuclear plant on March 29, 1979, not only happened at a time in which nuclear energy facilities construction faced rising costs and structural issues. It also occurred during a uniquely bad week in American history, when over 530 theaters across the country screened the blockbuster film *The China Syndrome*, released just 12 days before the accident.

The plot of the film depicted an accident at the fictional Ventana Nuclear Power Plant, much like Three Mile Island in size and location. The accident occurs because of faulty welds that allow a breakdown in the structural integrity of the facility. The film's title is drawn from the exaggerated fear that a core might melt through the Earth's mantle from Pennsylvania and reach China on the other side of the globe. The reactor's management takes draconian steps to prevent information reaching the public about the flaws that eventually caused the accident. Since many in the media and in the public had limited knowledge of nuclear technology, the film provided their sole exposure to the topic. The existence of the film set the stage for maximizing public fear and distrust regarding nuclear energy when Three Mile Island's partial core meltdown took place.

Three Mile Island

The Mile Island Power Station near Harrisburg, Pennsylvania had two pressurized-water reactors (PWRs) designed and built by American energy and technology service provider Babcock & Wilcox. Unit 1 (TMI-1), with an output of 819 MW electric, had been in operation since 1974 and had one of the best performance records in the United States until it was shut down in 2019. Unit 2 (TMI-2), with an output of 880 MW electric, was almost brand new at the time of the accident.

The events at Three Mile Island were caused by a combination of human error, equipment failure, and poor design of the human–machine interface. The accident started in the secondary loop, which was responsible for transferring heat to the steam generators. Subsequent actions led to a valve sticking open, which released cooling water. A routine maintenance operation had led to feedwater pumps turning off. The indicators on the control panel were misleading and caused the operators to turn off the pumps working to remedy overheating that had begun. The operators were not trained to recognize the ambiguity of some of the information they received. As a result, about half of the fuel melted before coolant was restored. The colder cooling water also shattered some of the hot fuel rods and all the fuel cells were damaged.

Progression of the Accident

Valve Malfunction

The principal mechanical cause of the accident at Three Mile Island was the malfunction of valves in TMI-2. Valve problems were not unique to the TMI-2 design and affected several other Babcock & Wilcox reactors, including the Davis–Besse reactor in Carroll Township, Ohio, which experienced a "near miss" accident in 1977. In this incident, a valve stuck open, after which, in a misreading of the reactor conditions, the operators shut off the emergency cooling system, causing reactor pressure and temperature to rise. Fortunately, the reactor happened to be operating in low-power mode, and within 22 minutes an operator correctly identified the stuck valve and took corrective measures. The incident sparked major concern among safety specialists at Babcock & Wilcox, whose warning memos nevertheless went unheeded as the company failed to update its training programs or inform its reactor customers. This left operators at similar plants in the dark about the potential for operator error arising from poor control room indicators leading to a loss-of-coolant accident. This is precisely what occurred at Three Mile Island two years later.

At the time of the accident, TMI-2 had only been in full commercial operation for three months. The previous nine months of testing had uncovered numerous issues in TMI-2 regarding valve operation, during which time the reactor automatically shut down, or "SCRAMMED," around 20 times. While operational challenges were not unusual during the testing period for new nuclear reactors, TMI-2 had experienced more problems than the average, resulting in the unit being non-operational for around 70% of the time.

On March 29, 1979, a fault in the TMI-2 feedwater purification system, which was undergoing maintenance, tripped the main feedwater pumps responsible for circulating water coolant over the reactor core. Within minutes, over 100 alarms blared in the control room. At this point, auxiliary feedwater pumps should have restored coolant flow, giving operators time to diagnose and correct the situation. However, a maintenance error had left critical valves shut. Thus, without coolant circulation, the temperature and pressure in the reactor rose, activating an automatic shutdown procedure that "SCRAMMED" the reactor, dropping in control rods.

After the SCRAM, decay heat continued to increase the pressure within the reactor. Some pressure fluctuations are expected during normal operation, so to manage these fluctuations, a pressurizer tank equipped with a pressure-operated release valve (PORV) creates a buffer. In this instance, as temperature and pressure rose, the PORV did half of its job: although it opened as it should have, it failed to close once normal pressure had been restored to the reactor system. The design of the controls offered no way for operators to directly monitor the position of the PORV, critically delaying their awareness that it was stuck open. Coolant rapidly escaped the reactor system through the PORV, draining into tanks in the basement of the containment building. Enough coolant escaped that the containment tanks overflowed and spilled radioactive water onto the floor of the containment building basement. From there, the water gathered in a sump and was pumped into backup waste storage tanks in the auxiliary building, which eventually overflowed as well. Unlike the containment building, the auxiliary building had vents to the outside, which created a pathway for radiation to leak into the atmosphere.

As the pressure in the reactor continued to drop, the automatic emergency backup cooling system turned on, which started to pump water into the reactor. This was "defense in depth"—the guiding safety principle of the nuclear sector—in action. Despite valve failures in the primary cooling system, redundancy limited the risk of a catastrophic accident, as the backup cooling system operated as intended. Had it not been for erroneous operator interventions after this point, the backup cooling system would have kept the reactor cool, buying operators time to identify the stuck-open PORV, as at Davis–Besse 13 months earlier.

Operator Error

Operator intervention, without clear signals in the control room, escalated the accident. Unaware that they were experiencing a loss of coolant, operators determined that the backup cooling systems posed a risk of overfilling the reactor system. If water levels got too high, the reactor and the coolant circulation system could go "solid," with excessive pressure causing the coolant system to burst. To check the water level in the reactor, operators looked at measurements in the pressurizer containing the PORV. Seeing water levels rise but blind to the malfunctioning valve, operators assumed that water levels in the reactor had normalized. Believing there was enough water in the system to keep the reactor cool, they shut off both pumps, one after the other. This was precisely the error the Babcock & Wilcox safety engineers had attempted to flag following the incident at Davis–Besse but was ignored.

With the coolant rapidly escaping from the reactor and the emergency system intentionally shut off, the system pressure continued to drop. Ninety minutes later, the pressure in the reactor was so low that the main coolant pumps began to vibrate violently. To prevent damage, operators began to shut those down. With the auxiliary pumps blocked, the primary pumps off, and the emergency cooling system manually shut down, there was no circulation of water to cool the core, and temperatures spiked. Operators suspected there was a leak at this point, but they first diagnosed it as one in the steam system instead of a loss of coolant in the primary loop. As water continued to flood the basement of the containment and auxiliary buildings, the water level in the reactor dropped to expose the reactor core, which quickly overheated. The zirconium rods that housed the fuel began to oxidize and fail, and the fuel began to melt. Radioactive gases escaped from the fuel rods and through the open relief valve into the auxiliary building and through ventilation into the outside world.

At 6:22 a.m., operators realized the release valve was stuck open, two hours and 22 minutes after the initial alarms. At this point, high temperatures and the oxidation of the zirconium fuel rods prevented a return to normal cooling. By 6:55 a.m. a site emergency was declared, and at 7:24 it was elevated to the status of general site emergency. At 8 a.m., site personnel identified that there had been some fuel damage, the extent unknown. The first news broadcast of the event took place at 8:25 a.m.

Aftermath

The series of events, which is described here in only moderate detail, provides an indication of the extreme complexity of these systems and the non-trivial ways in which humans and machines interface during their operation. Chaos and uncertainty marked the events, and as such a reconstruction of what went wrong was only possible through close study after the incident.

In response to initial broadcasts regarding the incident, public panic was minimal. A limited flow of information to the public was careful to not downplay the gravity of the incident. Radiation releases were reduced when Three Mile Island personnel stopped pumping the water from the basement of the containment building to the auxiliary building, where the ventilation stack was releasing a small amount of radioactive gas to the outside. In all, about 700,000 gallons of radioactive cooling water had traveled to the basement of the reactor building and into tanks in the auxiliary building.

The zirconium fuel rods continued to oxidize with the release of hydrogen, which accumulated to unstable levels within the containment building. At 1:50 p.m. a small hydrogen explosion occurred within the containment building but did not threaten its structure. Operators made progress in assessing and controlling the situation, and by the evening the reactor had cooled enough to resume normal functioning of the pumps. By the next day, March 29, the situation appeared to be stabilizing.

A couple of days after the accident some radioactive gas was released, but not enough to cause a dose to local residents that was above normal background levels. No injuries or adverse effects were experienced in the surrounding population. Further, there were no immediate injuries to the operators or staff of TMI-2. Studies into possible physical or medical problems that may have developed in the area around Three Mile Island all had negative results. The event did, however, have a gripping psychological effect.

On the morning of March 29, 1979, 28 hours after the accident, Pennsylvania Lieutenant Governor William Scranton III announced that the plant's owners had provided assurances that "everything was under control." This declaration set the stage for many contradictory statements to come as the situation unexpectedly escalated. It was later that day that Scranton had to modify his account and report that the "situation was more complex than the company had first led us to believe."

Two major events on March 30 entirely reversed the sense of control that the state government and plant managers had signaled. As operators depressurized the overflowed storage tanks in the auxiliary building, a burst of radiation from the plant ventilation system was picked up by a helicopter hovering above, which recorded a level of 1,200 millirem per hour—"many times the normal reading." Wrongly interpreted as a ground reading instead of a high-altitude reading that posed no public threat, and on the advice of the chairman of the NRC, Pennsylvania Governor Richard Thornburgh held an emergency press conference to instruct pregnant women and children within five miles to evacuate, which heightened levels of public fear.

Contingency plans were formulated for the city of Harrisburg to be evacuated if the situation worsened. These actions were taken despite no evidence of a radiological or explosive threat to public health.

The same day, a leading nuclear reactor physicist pronounced that a dangerous hydrogen bubble had formed inside the reactor and that it might explode, causing an uncontrolled release of radioactive material over a vast area. This assumption was soon found to be inaccurate, but not before prompting severe anxiety and leading to additional evacuations of around 140,000 of the 663,500 people living in the area.

The confused and contradictory management of the aftermath created a high-stress environment. Throughout the first week after the accident extensive speculation about the severity of the accident appeared in the press as reporters in both local and national newspapers struggled to cover the development of events that required sophisticated technical understanding.

President Carter Intervenes

Amid this chaotic situation, President Jimmy Carter, a nuclear engineering graduate of the United States Naval Academy, visited the site of the accident with his wife Rosalynn. After touring the contaminated control room, he issued a statement assuring "the people of this region that everything possible is being done and will be done to cope with these problems, both at the reactor and in the contingency planning for all eventualities that might occur in the future."

Between the accident on March 28 and the arrival of President Carter on April 1, there had been multiple announcements from authorities about the accident and its potential hazardous consequences. These announcements came from the reactor staff, from General Public Utilities (the utility conglomerate that owned the reactor), from officials associated with the State of Pennsylvania, and from the NRC. Some of these announcements were ill-founded and contradictory, which led to uncertainty and frustration in the public and the press. In response to this confusing array of pronouncements, Carter engaged Director of the NRC Office of Nuclear Reactor Regulation Harold Denton as his personal representative and coordinating spokesperson for all the agencies and offices involved with the accident. Denton had the personality, expert knowledge, and ability to communicate effectively, all of which helped to calm the turbulent atmosphere surrounding the incident.

One of the first issues Denton needed to address was the acute concern over the presence of a combustible hydrogen bubble within the reactor containment area. Denton briefed President Carter on this development minutes prior to Carter entering the plant to inspect the damage. For more than a day, the danger of a hydrogen-induced explosion generated calls for a general evacuation. The concern was dispelled when a second team of experts found that the first analysis had used the wrong formula and that, in fact, the release of hydrogen posed no risk of a large explosion.

Through his visit to the site, his personal review of the situation in the control room with Rosalynn, and his appointment of Denton as senior spokesperson, President Carter brought a calmer and more coherent approach to the chaotic discourse that had evolved. He was uniquely prepared to play that role via his experience in providing leadership in response to a meltdown of the NRX, a similar reactor in Canada.

Carter, after graduating from the United States Naval Academy in 1946, served in the Navy as a nuclear engineer. In 1952, he was working with a team in Schenectady,

New York that built the first nuclear submarine. That program received support on fuel studies from a reactor in Chalk River, Canada that experienced a partial meltdown in 1952. Carter led a team of 13 Navy volunteers who traveled to Canada to help in the cleanup. Carter and his team entered a highly radioactive area in the reactor in an operation that succeeded in removing contaminated components of the reactor core. They performed this task in highly orchestrated sequential operations lasting no longer than 90 seconds. For six months following this episode, Carter's urine tested positive for radioactivity. Hence, on the occasion of the Three Mile Island accident, President Carter was one of the few people in the United States who had had direct involvement with a nuclear meltdown.

President Carter's hands-on response quelled much of the public fear. After a week, headlines began to move on from the accident, which was by then fully under control.

The Kemeny Commission

Following the accident, President Carter established the Kemeny Commission to study the accident at Three Mile Island. A comprehensive report, *The Need for Change: The Legacy of TMI*, was published in October 1979, and revealed deep flaws in the organization and management of nuclear energy that went beyond the immediate events that occurred at Three Mile Island. The Kemeny Commission Report found problems in every component of the system responsible for nuclear reactor design, development, management, and operations. For Three Mile Island, this included the immediate staff responsible for operations, the utility company that owned the facility, the manufacturer who built the plant, and the agency that issued licenses for construction and operation of the reactor.

The Commission's report was unusually candid and insightful, due in large part to its unusual composition. The 12-member commission was chaired by President of Dartmouth College John G. Kemeny, a mathematician and computer scientist. Members included the head of a leading technology corporation, the president of an important labor union, the head of a key environmental organization, a professor of nuclear engineering, a leading nuclear scientist and inventor, and a housewife from the community adjacent to Three Mile Island. These individuals brought an unusually broad perspective to bear on the issues studied. They also were supported by a staff of experts and researchers.

The overall conclusion of the Kemeny Commission stated that "To prevent nuclear accidents as serious as Three Mile Island, fundamental changes will be necessary in the organization, procedures, and practices—above all—in the attitudes of the NRC and, to the extent that the institutions that we investigated are typical of the nuclear industry." Given that nuclear power had been an integral part of national infrastructure for more than 25 years, this was a damning report. The Kemeny Commission uncovered shockingly lax behavior and sheer mismanagement on the part of key players. A clear example of this behavior was the near-accident at Davis–Besse 13 months earlier, when operators had mistakenly shut off the emergency cooling system after the PORV malfunction. The apathetic response by Babcock & Wilcox management meant that other operators or members of the regulatory community were unable to learn from the mistake.

Beyond regulatory and technical problems, the Commission identified a fundamental issue that it categorized as the "mindset" behind operators assumed their actions were helpful without recognizing their role as a part of the problem. This blind spot meant that training for reactor operators was insufficient in much of the sector. This was exemplified by the Davis–Besse incident, as well as another in 1975 at the Browns Ferry nuclear plant in Tennessee, when the practice of checking for air leaks with a lit candle led to a highly publicized electrical fire.

The report also faulted an approach that focused on large-scale accidents rather than combinations of smaller incidents, such as those that occurred at Three Mile Island. It went on to emphasize that operators and supervisors must fully understand the plant's functioning and be able to respond to combinations of small equipment failures as well as large incidents.

The report also described flaws in the design of the control room in some detail. The accident resulted in more than 100 alarms going off in the control room, but in some cases, alarm signals were hidden by obstacles blocking a clear view. The Commission noted that "Overall, little attention had been paid to the interaction between human beings and machines under the rapidly changing and confusing circumstances of an accident."

However, the recipient of the most scathing comments from the Kemeny Commission was the NRC. The Commission argued that the culture and infrastructure underpinning nuclear energy generation was determined more by the policies and practices of the NRC than by actual operational experience in the nuclear environment. Rather than focus on operational learning and improvements, the report determined that the NRC equated safety with ever-increasing regulations, resulting in the development of voluminous and complex protocols that required an immense effort on the part of the facility and its employees to absorb into their operations.

The primary criticism of the NRC by the Commission was that it was so preoccupied by the minutiae of licensing that it lost sight of what should have been its primary concern: that of safety in all its dimensions. For example, the NRC categorized the analysis of low-probability accidents as a generic issue, yet it issued operating licenses without performing such analyses. It accumulated enormous amounts of information about plant operations without engaging in systematic review of that data for potential safety problems. They were criticized by the General Accounting Office for this behavior in 1978 but took no steps to address it before the Three Mile Island accident.

The Commission also noted that major offices within the NRC operated independently, with little evidence that they exchanged information or experiences. Lack of organizational structure meant that some issues that caused difficulty for operators were noted without steps taken to correct them. Moreover, the Commission observed that a key component of the NRC safety review was conducted by the Advisory Committee on Reactor Safeguards—a part-time group with inadequate staff and no firm guidelines or procedures for their review. Thus, safety concerns were relegated to an unstructured and inadequate review process.

The overriding concern of the Kemeny Commission was the way humans interacted with the systems that controlled reactor operations. They noted that

> There is no office within NRC that specifically examines the interface between machines and human beings. There seems to be a persistent assumption that plant safety is assured by engineering equipment, and a concomitant neglect of human beings who could defeat it if they do not have adequate training, operating procedures and controls.

This criticism of the NRC extended to the entire nuclear industry, including the manufacturers and the operating companies. The damning statement could be interpreted as a deep criticism of the engineering profession itself.

Building upon its extensive citation of inadequacies of the NRC, the Commission called for comprehensive restructuring of the agency, saying "The Commission believes that as presently constituted, the NRC does not possess the organizational and management capabilities necessary for the effective pursuit of safety goals. The Commission recommends that the NRC be restructured as an independent agency in the executive branch." They went on to identify many specific changes that were needed in the organization. While the Commission's recommendations did not lead to a total restructuring of the agency, it did bring about several reforms that strengthened its ability to improve its attention to safety considerations.

The Commission also had numerous recommendations addressed to the utility. High on that list was a call for the industry to " . . . establish a program that specifies appropriate safety standards including those for management, quality assurance, and operating procedures and practices, and that conducts independent evaluations."

Impact on Public Opinion

The Three Mile Island event was heavy ammunition for the anti-nuclear movement. The environmental lobby had achieved major wins without a concrete accident. Even the Brown's Ferry incident in 1975, in which the shared electrical system for a multi-unit nuclear plant caught fire and took out emergency cooling systems, ended with a safe automatic shutdown and no release of radiation. Now, a harrowing accident and measurable release of radiation had captivated the American public. The country was newly aware of nuclear energy, and not in a good way.

Public support for nuclear energy dropped below 50% after Three Mile Island and opposition surged into the mainstream. In May 1979, around 60,000 people marched on Washington, DC, chanting "No Nukes!" and rivaling the attendance at the July 1978 Equal Rights Amendment March. California Governor Jerry Brown spoke at the demonstration, calling nuclear power a "pathological addiction . . . storing up for generations to come evils and risks that the human mind can barely grasp." In September of that same year, a "No Nukes" concert at Madison Square Garden in New York City featured famous musicians including Crosby, Stills, and Nash, Bruce Springsteen, The Doobie Brothers, Jackson Browne, Bonnie Raitt, and others, attracting a crowd of nearly 200,000. The fervor of the anti-nuclear movement eventually waned, but nuclear energy remained publicly unpopular for decades.

Impact on the Nuclear Sector

The attitude toward nuclear power was permanently changed. In one decade, the sector had transitioned from bounding optimism to partial paralysis. If utility companies were afraid to build nuclear plants for economic reasons, they were now also afraid to do so amid overpowering public hostility.

A new wave of regulation after Three Mile Island dramatically increased the costs and timeline of the dozens of reactors under construction due to a two-part licensing/permitting process described in Part 50 of the NRC regulations. Here, nuclear operators were expected to comply with the regulations at the time their *operating* license was issued, which, given the pace of regulatory development, could be substantially different to those that existed years earlier at the original award of their *construction* permit. This led to large expenditure on construction with the potential for problems arising during the operational review. At that point the reviewers had a heavy burden to justify undoing completed construction. Part 52 regulations addressed this error, putting a "freeze" on regulatory requirements upon the issue of a site's construction permit, but not before substantial retrofits, blueprint changes, and reversed progress on reactors under construction following the Three Mile Island incident.

Reactors under construction at the time, such as the New Hampshire Seabrook Station Nuclear Power Plant, also confronted new gales of opposition. Originally welcomed by the town of Seabrook, the plant came to experience the full force of public opposition, including numerous lawsuits, political obstruction, acts of civil disobedience at the construction site, and refusal by the surrounding area to provide evacuation planning, thereby halting construction for a time. All of this made attracting capital for the project extremely difficult, ultimately driving the utility to the fourth-largest bankruptcy in US history at the time. Construction began in 1976, with two units planned, but by 1984, the second unit was canceled after nearly $1 billion in spending. The first unit, at last, came online 14 years and several times over the initial budget after the construction permit was granted.

Ongoing construction at other sites meant that nuclear capacity continued to enter operation throughout the 1980s, but new reactor orders ceased entirely. However, the sector was far from dormant. It had turned inward and, in the wake of the Three Mile Island accident, was implementing major reforms that created positive trends in operation and safety, which have endured in the US reactor fleet since, including increased uptime and decreased instances of events requiring remedial actions. The NRC implemented major reforms in inspections, agency staffing and capacity, emergency preparedness, and other key areas.

Impact on Course of American Energy Production

Ambitious projections by the AEC (which became the Department of Energy) that the United States would get around 55% of its energy from nuclear power by 2000 were never realized. In place of canceled reactors, the United States burned incredible amounts of coal, with vast total emissions and health impacts from the particulate pollution and carbon dioxide release.

It is hard not to speculate how greatly US emissions could have been reduced, and how much cumulative learning and improvements on nuclear energy could have been accomplished, had the United States not abandoned new nuclear construction after Three Mile Island. However, it is simplistic to ascribe the nuclear energy bust to one unlucky week in March 1979. The nuclear sector was already in trouble and needed technical, financial, organizational, regulatory, and communication innovations. Three Mile Island was a reckoning that the nuclear sector did not want but perhaps needed.

The accident and the ensuing investigations pushed the nuclear sector toward higher standards of safety and operation. Unfortunately, the event was sensationalized by antinuclear forces who exaggerated its impact, which pushed fears and paranoia to higher levels. Had the events been put into accurate perspective a more balanced outlook on nuclear energy might have emerged.

Further Reading

Report of the President's Commission on the Accident at Three Mile Island: The Need for Change: The Legacy of TMI, Pergamon Press, 1979.
Known as the Kemeny Commission Report, this historic document recounts the events of the accident and the Commission's recommendations for needed changes and policies of the Nuclear Regulatory Commission, as well as needed modifications in nuclear power plant management by a cognizant utility.
J. Samuel Walker, *Three Mile Island: A Nuclear Crisis in Historical Perspective*, University of California Press, 2006.
Historian J. Samuel Walker describes the Three Mile Island accident in the context of the contemporary nuclear-related discourse and analyzes related social, technical, and political issues.
Natasha Zaretsky, *Radiation Nation: Three Mile Island and the Political Transformation of the 1970s*, Columbia University Press, 2018.
An insightful analysis of the accident and its impact on American culture.
Thomas H. Moss and David L. Sills, eds., *The Three Mile Island Accident: Lessons and Implications*, New York Academy of Sciences, 1981.
A collection of papers from a conference held by the Academy on April 24, 1981.

12

The Chernobyl Accident

As you all know, a misfortune has befallen us—the accident at the Chernobyl nuclear power plant. It has painfully affected Soviet people and caused the anxiety of the international public. For the first time ever we encountered in reality such a sinister force as nuclear energy that has escaped control. So what did happen?

— *Mikhail Gorbachev*[1]

Chernobyl Overview

The explosions that destroyed reactor number 4 at Chernobyl, Ukraine on April 26, 1986, still reverberate today as the possibilities of using nuclear energy to meet the challenges of a growing world population and global warming are assessed. The triad of nuclear accidents that have had profound consequences are Three Mile Island, Chernobyl, and Fukushima. Of the three, Chernobyl had, and continues to have, the greatest impact on society's perception of nuclear energy. Nearly 40 years after the event, society is still grappling with the causes of Chernobyl's demise and the lessons that can be learned from that catastrophe. If those causes were inextricably interwoven with fundamental design features of nuclear power plants, then that technology might be forever eschewed. However, if Chernobyl ultimately is judged to be an idiosyncratic event, nuclear energy development and use may continue to grow.

The history of the decision to build this reactor, the events surrounding its construction, the details regarding the events preceding and after the tragic explosions, and the way this calamitous event was presented to the people of the Soviet Union and then to the world are clouded by multiple instances of obfuscation, misrepresentation, and cover-up. To unravel this muddled history, this chapter attempts to confront the myriad questionable actions and outright lies that one confronts in a systematic manner in search of lessons that might be drawn from this catastrophe.

The stage was set for eventual problems when plans were first formulated for the design and construction of the Chernobyl reactor in the late 1970s. At that time,

---

[1] Mikhail Gorbachev, "Excerpts from Gorbachev's Speech on Chernobyl Accident", *The New York Times*, May 14, 1986. https://www.nytimes.com/1986/05/15/world/excerpts-from-gorbachev-s-speech-on-chernobyl-accident.html, accessed January 30, 2025.

Nuclear Energy. Edward A. Friedman, Oxford University Press. © Edward A. Friedman (2025). DOI: 10.1093/9780198925811.003.0012

decisions were made about the development of nuclear power by senior bureaucrats whose deliberations were largely secret and unvetted by scientists and engineers. Warnings dating from 1965 from both scientists and engineers that the design of the Chernobyl reactor was dangerously flawed were ignored.

It was clear that this reactor should not have been built. This type of reactor was designed by the military to optimize the production of plutonium for nuclear weapons. It was decided early on by Soviet bureaucrats to use this military reactor for civilian applications. It was a light-water-cooled graphite-moderated reactor with 2% enriched uranium and boron–carbide control rods. During normal operations, some of the uranium 238 (U-238) in the fuel rods absorbs a neutron and transforms into plutonium 239 (Pu-239), which is highly desirable as the explosive component of a nuclear weapon.

There were 1,600 fuel elements that could be extracted individually with the use of a crane that was positioned on top of the reactor (Figure 12.1). That arrangement allowed the extraction of fuel elements at the most desirable moment for harvesting of weapons-grade plutonium. This reactor design, which was optimized for production of weapons-grade plutonium, did not consider the safety features required of civilian power reactors then being developed in the United States, Canada, and elsewhere.

Figure 12.1 *The reactor lid for Unit 2, Chernobyl Nuclear Power Plant.*
Reproduced courtesy of Carl Willis (Albuquerque, NM, USA).

Design Flaws

This reactor, known in the Soviet Union as the RBMK, had several major design flaws, one of which was that the facility had no containment enclosure. Containment enclosures were used routinely elsewhere in the world to prevent material from a steam or hydrogen gas explosion from contaminating the environment. Because this reactor design had been used in the Soviet Union in the 1970s without serious problems, a false sense of confidence prevailed at the time of its introduction for civilian power in 1983. While many REMK reactors were built in the Soviet Union and in Soviet satellite countries, no other country in the world used the RBMK design.

A second major design flaw was a reaction to overheating that triggered more over-heating and which proved fatal to the reactor. As early as the mid-1960s bureaucrats knew about this problem, which was a result of the water that flowed through the fuel rods as a coolant also absorbing neutrons, thus limiting the intensity of the heat-producing chain reaction.

Scientists had pointed out that when this water became overheated, steam bubbles would form, creating voids that would allow an increased flux of neutrons, thus causing more heat to be generated. This sequence of increased heating, which causes even more heating (known as a positive feedback loop) can easily get out of control—and it did! The critique from the 1960s was not only ignored, but it was also suppressed.

A third major design flaw was a consequence of a rush to complete installation ahead of schedule, to receive bonuses, thereby neglecting efforts to install a fully operational backup power system. The reactor went online on December 20, 1983, two days before the completion deadline, thus qualifying employees to receive awards and bonuses for good work in completing construction early.

In accelerating the reactor's opening, the construction team skipped installing a very important safety mechanism for the backup emergency cooling to compensate for a power failure in the region. The facility's pumps that operated the cooling system for the fuel rods would shut down at the power failure, but within a minute, emergency generators would provide the electricity for cooling, leaving only 60 seconds without electricity. Establishing an emergency electrical system was needed to maintain operations during that critical one-minute period. In the rush to go online in time to qualify for awards and bonuses, this task was deferred.

When reactor four at Chernobyl was placed in operation, there was no fully functional emergency electrical system in place to meet the need for the first minute of regional power loss. This defect was kept secret within the operational staff of the reactor and was a source of constant anxiety. The disaster occurred when on April 26, 1986, the long-delayed effort was made to test an electrical system designed to cover that one minute. The botched implementation of that effort was the direct cause of the following disaster.

A fourth major design flaw exacerbated the problems that arose from the overheating of the reactor's core. Amazingly, the control rods, intended to reduce the rate of fission, instead increased it. The control rods were made of boron–carbide, a material that absorbs neutrons. However, the tips of these control rods were made of graphite, which did not absorb neutrons but displaced water that had been absorbing them, increasing the neutron flux. Thus, when the operators sought to curtail the overheating of the reactor by inserting the control rods, they caused the heating to intensify.

Undisclosed Pre-accident Information

The history of the Soviet Union is replete with examples of events and government actions that have been secret from the public. Further, there are many cases where information about nuclear accidents were withheld not only from the public, but also from those holding key positions in the field of nuclear energy itself. When open discussion of problems and accidents is suppressed, it is impossible to nurture a safety-oriented culture. This absence of a safety culture in the era prior to the Chernobyl disaster allowed misjudgments and mistakes that were counterproductive, if not dangerous.

A notable example of this behavior was an event of similar magnitude that took place in 1957 in the Ural Mountains of the USSR and which pre-dated those of Three Mile Island, Windscale, Chernobyl, and Fukushima. While not a reactor accident per se, it was an event intimately connected to nuclear reactor technology that should have been known to all nuclear engineers in the Soviet Union. It occurred near Kyshtym, where the Soviets were developing nuclear weapons and had large nuclear waste accumulations, one of which exploded. It was a chemical explosion in the waste containment area, rather than a nuclear explosion and was estimated to have been equivalent to that of about 85 tons of TNT. While minor compared to the Hiroshima bomb explosion (equivalent to about 15,000 tons of TNT), the impact was considerable, requiring an evacuation of more than 10,000 people within an area larger than that of New York City that became seriously contaminated. While the number of associated deaths is not known, claims that there were more than 300 immediate deaths among the residents of villages in the area are plausible. Since the area was remote and world monitoring of radioactive fallout was limited, the Soviets were able to keep the event secret until 1981. As a result, nuclear engineers and others working in the nuclear industry in the Soviet Union did not have the opportunity to learn about radiation, its effects, and all the other related issues associated with a major nuclear accident.

In the 1970s, there were Chernobyl-like RBMK reactors in the Soviet Union, with graphite moderating structures holding many individual fuel assemblies. These reactors experienced various types of accidents: small explosions, fires, emergency shutdowns, which were reported neither in the Soviet Union nor to the international agencies. As a result of this secrecy, professionals in the industry, especially operators and managers of RBMK nuclear power plants, were unable to learn anything from those experiences.

Secrecy Prior to Accident

Secrecy about the test of an emergency power system to protect the Chernobyl-4 reactor during the first minute of an external loss of power was a direct result of secrecy surrounding the failure to implement such a system when the plant opened in 1983.

Hence, when tests for such a system were planned in 1986, details of this initiative were known only to a limited number of administrators and operators at the plant and to no one in the regional power infrastructure.

After the reactor opened without full testing, there was an urgent need to implement the missed test of an emergency electrical system. In 1984, such a test failed, and in 1985 another trial emergency mechanism also failed. The cognizant managers at the reactor were nervous and knowledge of the missing component to the backup electrical system was not shared by plant management with higher-level bureaucrats in the nuclear management and government power hierarchy. The Chernobyl-4 reactor came online following prior installation of three other RBMK reactors, with two additional reactors of this design waiting for installation. Not only was the integrity of Chernobyl-4 consequential, but the viability of the RBMK design was of critical importance for national energy planning. This location in Ukraine was envisioned by central planners in the Soviet Union to become the site of the largest concentration of nuclear power plants in the world.

Unfortunately, at the time of the disastrous test on April 26, 1986, this reactor had been in operation for quite a while, during which time the fuel being used underwent fission reactions that created copious amounts of radioactive byproducts. At the time of the test, the fuel containers had produced substantial amounts of dangerous radioactive material. Thus, it was the worst possible time for conducting this test.

Information about the test was kept from the regional energy management personnel, who took action to reduce power levels during the day of April 25. Unfortunately, this powering down corresponded exactly to the time scheduled for the test for which the daytime crew had been trained to conduct the operation. The test was postponed to after midnight on April 26, when a crew would be on duty that had no prior knowledge of what was planned.

The Dysfunctional Test Triggers Explosions

On April 25, the power reduction mandated by the regional power authority set the stage for the disaster. The precipitating factor was the high concentration of ^{135}Xe, which is a robust absorber of neutrons. When the reactor power was lowered it continued to build up the fission product ^{135}I, which decays into ^{135}Xe, and which, in high concentrations, impedes action to increase the power level of a reactor. The xenon concentration is depleted by natural radioactive decay and by absorption of neutrons. At low power levels the neutron flux is low and the continued production of ^{135}Xe from the decay of ^{135}I exceeds the rate of depletion. Given that the half-life of ^{135}I is eight days, the production of ^{135}Xe increases for about ten hours after a restart of a reactor that had been in a near shut-off condition. This phenomenon creates a natural brake on efforts to restart a reactor. It is only after ten hours that the neutron flux is sufficiently high and the production of new ^{135}Xe from residual ^{35}I is sufficiently low enough for the reactor to sustain a significant power increase.

This inhibition of start-up due to absorption of neutrons by ^{135}Xe is known as xenon poisoning. This condition, which was brought about by the reduced power levels, made it impossible to restart the reactor in the time for which the test was rescheduled. However, the managers and operators on duty during the early morning of April 26 were not familiar with xenon poisoning and tried to initiate what was a hopeless effort to quickly restart the reactor.

Faced with a failed restart, the operators removed many of the control rods and shut off emergency systems. Eventually, the boiling process that precipitates positive feedback kicked in. The voids in the boiling water led to more neutrons initiating additional heat-producing fission with an exponential increase in temperature of the fuel rods. At this point, the operators realized that it was essential to shut down the reactor. The operators tried to insert all the available control rods, not realizing that the control rod design was flawed. When the tips of the control rods were inserted, the result was an increased power level. The tips of the control rods displaced water-absorbed neutrons with graphite, allowing a larger flux of neutrons and an increase in power production.

The runaway heating caused not only the melting of fuel rods but also the oxidation of the titanium fuel rod metal containers. As oxygen combined with titanium released hydrogen from the overheated water molecules, the hydrogen accumulated and caused explosions that, along with a steam explosion, ripped the reactor apart. It must be stated explicitly that these destructive explosions were caused by steam and hydrogen, not by nuclear fission.

The largest explosion took place at 1:23 a.m. on April 26, 1986. There was a 2,000-ton cover over the reactor but no containment structure. This vast cover became vertical and broke apart. The large graphite matrix of the reactor is a form of carbon used for fires. Thus, the exposed graphite body of the reactor ignited. A great deal of radioactive material and some uranium and plutonium flew out of the reactor. The fire continued for some time before it could be stopped. Extensive amounts of radioactive material were dispersed over regions more than 2,000 miles distant. For about two weeks after the explosion, the weather patterns sent plumes of radioactive material in all directions.

Secrecy and Dysfunction after the Explosions

At first, the Soviet government tried to keep the explosions secret but this was impossible. On April 28, 1986, workers at a nuclear power plant in Sweden thought that they had a problem at their facility because they were measuring serious levels of radioactivity. They soon realized that it was coming from someplace else. They analyzed airborne particles that were coming through to Sweden and found isotopes that could only have originated in a nuclear reactor with melted fuel rods. The clear conclusion was that a reactor had failed in Ukraine. They then confronted the Soviets with these findings. After initial denials, the Soviets finally admitted that there had been an accident. At 9 p.m. on April 28, there was a report on Moscow television. It was a succinct report which stated that, "An accident has occurred at the Chernobyl Power Plant. One of the reactors has been damaged. Measures are being taken to eliminate the consequences of the accident. Aid is being given to those affected. A government commission has been set up."

After the explosion, other gaps in knowledge became apparent. Most tragically, the firefighters and other initial responders didn't know what they were confronting. They lacked basic understanding of radioactivity, and they did not have instruments for its measurement. Also, they did not have protective clothing nor knew what precautions were needed. Consequently, the death rate among firefighters was high.

Nearby to the Chernobyl-4 reactor, the nuclear workers town of Pripyat, with a population of 50,000 people, was not immediately told that there had been a serious reactor explosion with elevated levels of radioactivity in the region. The people of Pripyat, unaware of the dangers to which they were exposed, went about their daily activities as usual. Normal events were held, children played outside, people strolled in the streets: this behavior should not have been allowed. Since news of the accident was not disclosed until 9 p.m. on the night of April 28, secrecy prevented proper action being taken in Pripyat. This secrecy also seriously delayed planning actions needed to protect citizens being exposed to high radiation levels. These delays resulted in the people of Pripyat not being evacuated until 2 p.m. on April 28. More than 1,000 buses succeeded in evacuating the population that afternoon.

While in the region near to the damaged reactor, protective action was delayed, during the ensuing days, in more distant regions, needed action was ignored. Kyiv, which is just 63 miles from Chernobyl, was then the eighth largest city in Europe with a population of 2.5 million people. The residents of Kyiv were not alerted to the problem and went about their business as usual for weeks. In Kyiv and other cities in Ukraine and Russia, May Day was celebrated as usual, even though there were serious levels of radiation and fallout. Children of Kyiv were not evacuated until mid-May. The full health consequences of this irresponsible behavior by the Soviet authorities will never be known.

The world was quite agitated about the explosion and the aftermath of the Chernobyl disaster. There was intense concern throughout Europe and elsewhere. The International Atomic Energy Agency (IAEA) was under pressure to provide information and answers to pressing questions. The IAEA held an international conference in August of 1986 where a presentation from the at the Soviet Union was made by Academician Valery Legasov. Legasov, who had been instrumental in ameliorating the post-accident impact at the reactor site and nearby, presented his report in conformity with the instructions that he had received from the Politburo to lay the blame for the entire disaster on those who were working at the reactor. He asserted to the international community that managers and operators at the reactor had failed in their duties.

In keeping with the idea that everything that went wrong was due to the people managing and working at the plant itself, a show trial was held in July 1987 where six individuals who had been managers and operators were charged with negligence, irresponsibility, and mistakes in behavior. After being convicted, they were sentenced to between two and ten years in corrective labor camps. This was the last show trial that was held in the Soviet Union.

Not long after the 1986 IAEA conference, Legasov regretted what he had done. He became depressed and after about a year started writing his memoirs with a more accurate account of the events leading to the disaster. On the second anniversary of the Chernobyl disaster, he was found hanging in the hallway of his apartment house. While officially judged to be a suicide, there was strong speculation about other causes for his death. His memoirs were never completed.

Latest information about the Chernobyl disaster started to emerge in 1992 after the collapse of the Soviet Union. Misrepresentations of events presented at the 1986 IAEA conference were exposed. A second IAEA conference was held that year that produced a report asserting that the USSR lacked a culture of safety.

After the disaster, the three other reactors at Chernobyl were closed. It took some time for that to be implemented. It was not until 1991, 1996, and 2000 that the other three reactors ceased operations.

The unique RBMK design was implemented in Soviet satellite countries and elsewhere in the Soviet Union. The Soviets built two RBMK reactors in Lithuania with the result that Lithuania was not allowed accession into the European Union until those reactors were closed. That was done with considerable international support in 2000 and in 2009.

There are eight RBMK reactors that continue to operate in the Russian Federation. Russia says that these reactors have been modified and enhanced and that safety concerns have been addressed. However, these reactors would not be allowed to operate in Europe or the United States. Closing these reactors in the Russian Federation would be economically difficult since they provide more than 25% of nuclear generated electricity in the country. While it would be desirable for the IAEA to visit and analyze the safety of these reactors, there is little likelihood of that happening.

Zhores Medvedev

This chapter relies heavily on Zhores Medvedev's *The Legacy of Chernobyl* (see Further Reading). While there are many published accounts regarding the origin and consequences of the Chernobyl disaster, this author considers the that Medvedev's observations provide the most reliable available account, especially after a review of his impeccable credentials. Because there was both secrecy and cover-up regarding the origins and events leading up to and following the explosions on April 26, 1986, construction of an accurate narrative is challenging. Dr. Zhores Medvedev (November 14, 1925–November 15, 2018) was an accomplished biologist who held the rank of Senior Research Scientist. He was head of a prominent molecular radio-biology laboratory in the Soviet Union. In 1962, he wrote *The Rise and Fall of T. D. Lysenko*, an expose of the fallacies of Soviet genetics. It was published in the United States in 1969 by Columbia University Press and resulted in his dismissal from his academic positions in the Soviet Union.

In 1970, Soviet authorities began an attempt to have Dr. Medvedev isolated in an asylum for the insane. With help from his brother (also a distinguished scientist), support from human rights advocates inside the Soviet Union, and backing from the international community, the effort to have him committed failed. To expose his oppression by the Soviet government, Medvedev and his brother then wrote *A Question of Madness*, which was published in the United States in 1971. A year later, during a stay as a visiting scholar in England, he was stripped of his Soviet citizenship.

Dr. Medvedev continued to speak out for truth in science by publishing the first analysis of the 1957 Kyshtym explosion in his book *Nuclear Disaster in the Urals*, published in the United States in 1979. Mikhail Gorbachev restored his Soviet citizenship in 1990. In 2007 Medvedev published a series of articles linking the Polonium-210 poisoning of Alexander Litvinenko in London to the leadership of the Russian Federation. Considering the outstanding record of revealing nefarious actions on the part of his country's leadership, it is surprising that some scholars of the history of Chernobyl overlook the work of Zhores Medvedev.

Deaths Attributable to Chernobyl

Conflicting points of view arise due to differing assessments both of the radiation that resulted from the Chernobyl explosion as well as the impact of that radiation on the health of the exposed populations. Discussion of these issues requires some understanding of the physical origin of that radiation, as well as the biological consequences of radiation exposure.

The radiation that spread throughout the Northern Hemisphere because of the Chernobyl explosion had its origin in the nuclear fission process that produced the reactor's energy. As the uranium nuclei split, lighter atoms are produced, including many that are unstable. That instability leads to the emission of gamma rays (electromagnetic radiation with higher energy than X-rays), electrons, neutrons, and alpha particles (two neutrons and two protons). An alpha particle is identical to the nucleus of the most common type of helium atom. These various emissions can damage biological cells. The questions that arise ask how much radiation is produced and what are its consequences?

Conflicting Accounts

On the day of Chernobyl explosion, pump operator Valery Khodemchuk and systems operator Vladimir Shashenok died before the end of the day from the radiation and trauma of the explosion. By the end of May 1986, 23 other workers from the plant, as well as firefighters, died of acute radiation syndrome. These deaths from the immediate aftermath of the explosion are well documented. However, more than 30 years later, the question of how many subsequent deaths from radiation are attributable to the release of radioactivity from that disaster remains a matter of controversy and debate.

The Chernobyl Forum, a team of more than 100 experts assembled by eight UN-related agencies, asserted in 2005 that long-term consequences of Chernobyl could result in as many as 4,000 excess cancer deaths. Originally published September 5, 2005,

Chernobyl's Legacy: Health, Environmental and Socio-economic Impacts confines their analysis to the contaminated regions of Ukraine, Belarus, and Russia. While recognizing the increased radiation exposure throughout Europe, the report asserts that these levels were too small to cause an observable impact on the number of deaths due to cancer. Their logic is based on the fact that with approximately 20% of the population dying from cancer, excess deaths in the hundreds or thousands cannot be distinguished from the steady-state deaths that occur numbering in the millions.

While additional deaths due to radiation may not be directly observable, it does not mean that they have not (and do not) continued to occur. During the past 30 years, many scientists, government organizations, public interest groups, members of the press, and others have made predictions, speculated upon, and debated this issue. The number of immediate and long-term excess deaths in the world resulting from the Chernobyl disaster encapsulates in a single number a summary of the total devastating impact of what was the world's most catastrophic nuclear power plant accident. As such, it acts as a reference point that deeply influences attitudes toward nuclear power. Perhaps the single greatest factor influencing Germany's decision to eschew nuclear power is the enormous number of deaths from Chernobyl claimed by Greenpeace as possibly exceeding 200,000. Such a large number rests heavily on the psyche.

Other published figures range from 26,000 (by Lisbeth Gronlund of the Union of Concerned Scientists) and similar estimates by noted physicist Richard Garwin and the highly unlikely number of 985,000 (which appears in an English-language book by Russian scientists published by the New York Academy of Sciences).

Unsurprisingly, those who support nuclear power development are more likely to quote the lower numbers, while advocates seeking the elimination of nuclear power focus on the larger figures. Given the centrality of nuclear power in strategies that might help offset global warming, this chapter suggests that clarity on this number, which is so central to thinking about nuclear safety, is a major public policy issue that deserves scrutiny using the best available and relevant scientific understanding.

The Chemical Makeup of Chernobyl Fallout

To discuss radioactivity and fission, it is necessary to differentiate between the different forms of chemical elements according to the composition of their nuclei. Every element has a distinct chemistry and gains its identity by the number of protons in its nucleus. Hence all forms of hydrogen have one proton, and all forms of uranium have 92 protons. The chemistry of elements is determined by the number of electrons that surround the nucleus which is equal to the number of protons in the nucleus. However, elements can have varying numbers of neutrons. Hydrogen, for example, can have zero, one, or two neutrons. Chemically, these forms of hydrogen are identical since their atoms all have one electron. However, their nuclear properties differ. The form with one neutron is deuterium, while the form with two neutrons is tritium. In the case of uranium, the most common form of uranium has 146 neutrons, while the uranium used in atomic weapons has 143 neutrons.

A convenient way to account for these different forms of nuclei for the same chemical element is to add the number of neutrons to the number of protons and identify the result using the total. Hence deuterium can be called hydrogen-2 and tritium is known as hydrogen-3. These total numbers are designated as the isotope number. While the word "isotope" can be intimidating to a lay person with limited knowledge of science, it should be kept in mind that it is a bookkeeping number much like labeling a bag of groceries as "fruit fifteen" if it contains ten oranges and five apples. Using this nomenclature, we can add the 92 protons and 146 neutrons in atomic bomb uranium to get an isotope number of 235, while the most common natural uranium has an isotope number of 238.

The health consequences of the Chernobyl disaster are primarily due to exposure to radioactive substances and the inhalation or ingestion of radioactive substances. These radioactive materials were created in the nuclear reactor as a byproduct of the fission or breakup of uranium 235 (U-235) that produced the electrical energy used in homes and factories. The radioactivity of the original U-235 fuel is benign. However, when fission takes place, new atoms are created. Many different outcomes are possible. These outcomes of the splitting of U-235 have a statistical distribution that is predictable.

Some combinations resulting from the fission process are more likely than others. Of these many outcomes, only around a dozen have significant consequences concerning radiation exposure. All of these fission products are accompanied by radiation—gamma rays, alpha particles, beta particles (high-energy electrons), and neutrons.

Among the fission products, there are a few that present a danger to human beings through entrance into the digestive system via ingestion of milk and other foods. The most dangerous of these are ^{131}I, cesium-137, and strontium-90. Iodine is absorbed by the thyroid gland and can cause thyroid cancer if enough radioactive iodine is retained. Cesium and strontium enter the food chain and, if large enough amounts of the radioactive forms of these elements are present in the body, can cause other types of cancer. The chemistry of cesium is like that of potassium, and so is actively absorbed by the human body, while the chemistry of calcium is like that of strontium, which leads to its similar biological absorption by humans. The health consequences of radioactive materials are most severe when the active elements are ingested and emit their gamma rays, electrons, neutrons, or alpha particles inside the body, rather than from an external location. However, when the radioactive material enters the food chain, it emits radiation that adds to the potentially harmful dose received by inhabitants of the contaminated region.

Animals with High Levels of Radiation in Distant Locations

Given that the Chernobyl Forum includes prestigious organizations such as the World Health Organization (WHO) and the International Atomic Energy Agency (IAEA), it is clear that the Forum should not have neglected deaths outside of Russia, Ukraine, and Belarus. There is incontrovertible evidence for life-impairing levels of radioactive fallout at distances of more than 1,500 miles from the explosion in measurements of contaminated meat from sheep and reindeer in Scotland and Lapland.

In 1986, the Scottish government placed restrictions on 2,900 farms stocking 1.5 million sheep. In 1987, additional restrictions were imposed when it was discovered that the new season's lambs were highly contaminated. Significant restrictions were not lifted until 1991. As of 2008, five farms were still designated as restricted, and all monitoring was not ended until June 21, 2010.

In July 1993, the UK The Parliamentary Office of Science and Technology issued Briefing Note 45, which provided to members of Parliament information concerning Chernobyl fallout, and showed that the most significant levels of fallout were in northern Scotland, with additional levels of concern in west-central Scotland, Cumbria, Wales, and Northern Ireland. It also documents the number of sheep subject to inspection: Wales (2.1 million), Scotland (1.36 million), and England (0.87 million). Briefing Note 45 further identifies the percentage of those examined whose meat needed to be removed from the market: 22% in 1987, 10% in 1988, and 7% in 1989, with lower levels in subsequent years. These figures lead to the conclusion that meat from more than a million sheep was removed from market in the United Kingdom because of Chernobyl fallout. However, note that the impact of fallout from Chernobyl on the sheep of the United Kingdom is masked in later years by the occurrence of a huge outbreak of foot and mouth disease in 2001, which resulted in the slaughter of between seven and ten million sheep and cattle.

Cesium-137, which can easily enter the food chain, emits harmful gamma rays as well as electrons (called beta radiation). In nature, cesium has a half-life of about 30 years, that is, after 30 years half of the original deposit of cesium-137 will have undergone radioactive decay. After 60 years the level will be 25% of the original and after 90 years, the level will be one-eighth the original, and so on. Given the long half-life of cesium-137, it is unsurprising that restrictions and examinations of sheep in the United Kingdom lasted for 26 years, until 2010, when the controls were finally lifted.

While sheep farming was severely affected in Scotland, fallout from Chernobyl also contaminated thousands of reindeer in Lapland, which upset long-standing social and cultural patterns of the Saami people of the region. As one observer put it, the Chernobyl disaster, "scarred" their way of life. There are about 80,000 Saami: 50,000 in Norway, 20,000 in Sweden, 8,000 in Finland, and 2,000 in Russia. While reindeer meat is a staple of the Saami diet, the meat, organs, and other components of the reindeer play roles in their animistic religious practices. The Saami use the entirety of the reindeer body, from the entrails and organs to the antlers, hooves and blood.

Nearly 80% of the reindeer meat in Sweden was destroyed in the 1986 slaughter season. This severe economic loss led to the governments of Sweden and Norway providing compensation to the affected communities. However, the cultural uses of reindeer carcasses could not be replaced with monetary payments. A member of the Saami community summed up this loss as follows: "This is not just a matter of economics but of who we are, how we live, how we are connected to our deer and each other."

The impact of Chernobyl in this region has been long lasting. The contamination's severity was exacerbated by the Scandinavian reindeer consuming lichen as their main winter staple. Lichen does not have a root system and absorbs nutrients directly from the air, thus soaking up large concentrations of radioactive cesium. As recently as 2014, there were regions in Lapland where radiation levels in reindeer meat still exceeded safe levels as a result of the long-term stability of the lichen food stock and the 30-year half-life of cesium-137. In 2016, the fallout that remains in the environment still releases half of the radiation intensity it did 30 years earlier directly after the Chernobyl explosion.

Given that the distances from Chernobyl to London and Paris are less than the distances to the highlands of Scotland and the fields of Lapland, it is arguable that significant amounts of harmful radioactive material were deposited in areas of Western Europe in April and May of 1986. Chernobyl emitted radioactivity for approximately a week, during which time changing wind patterns carried it over a wide geographic region. Fallout was then concentrated via rainfall in specific locations, in some of which there was careful monitoring of the fallout. In other regions, most notably France, little attention was paid to this hazard. While most countries issued warnings about food that might be contaminated, such action was not taken immediately in France. Following the Chernobyl accident, France's Central Service of Protection against Ionizing Radiation (SCPRI) initially denied that the radioactive cloud had passed over France. This lack of action became a highly contentious issue in French politics. Public warnings were not issued in France about the hazards of consuming potentially contaminated milk and produce, thus adding to the difficulty of evaluating the long-term health impacts resulting from Chernobyl.

Lisbeth Gronlund's Analysis

Of the many analyses of the cancer deaths from Chernobyl, that of Lisbeth Gronlund, a senior scientist of the Union of Concerned Scientists, stands out as particularly well-reasoned and comprehensive. The data and scientific concepts on which her report April 27, 2011, are based have proven to be consistent with later studies and scientific research.

Gronlund specifically rejects the assertion of the Chernobyl Forum that the impact in regions beyond the high radiation fallout locations of Russia, Ukraine, and Belarus are too small to consider, and suggests that ". . . by limiting its analysis to people with the greatest exposure to released radiation, the report seriously underestimates the number of cancers and cancer deaths attributable to Chernobyl. The effects of the radiation were not limited to the contaminated areas but would be felt in Europe and beyond." Gronlund explicitly notes that areas of Asia, Africa, and the Americas were contaminated by the Chernobyl accident.

Gronlund uses results of international studies to estimate the radiation doses received in all the affected areas. It is assumed that the increased levels of radiation are due mainly to the fallout of cesium-137—the same isotope that caused such havoc with the sheep and reindeer herds. This cumulative dose is then multiplied by the probability of cancer deaths due to the received dose.

This chapter equates a dose of 2 millisieverts (mSv) to be equal to the average background radiation (B). Since a mSv is one thousandth of a sievert, a one-sievert dose is equal to 500 B doses. The probability of cancer death per sievert of received dose used by Gronlund is 5.7%.

Her calculations result in 4,000 deaths in the contaminated areas. To this total, which is consistent with the Forum evaluation, she estimates another 26,000 deaths: 4,000 among recovery operation workers, 5,000 from less-contaminated areas of the former Soviet Union, 9,000 from other European countries, and 4,000 from other northern hemisphere locations outside of Europe.

Given that there is considerable uncertainty in the 5.7% death-rate figure, Gronlund identifies the range of deaths that would result from using a 95% confidence interval, rather than a specific rate. Her calculation for the upper and lower bounds that fall within that confidence interval is 12,000–57,000. Note that these excess deaths do no include excess thyroid cancer deaths. While there were many thyroid cancers, the condition is treatable, and the number of deaths is minor compared to the figures for solid cancer and leukemia.

While Gronlund considers the uncertainties inherent in the calculation of excess deaths due to low-level radiation, there are other uncertainties that are more difficult to assess. For example, we do not know the extent to which flocks of sheep or agricultural produce in France were contaminated. These are known unknowns. It may also be the case that water resources—particularly in Ukraine—were contaminated and never documented. These can truly be said to be unknown unknowns. Radioactive elements that are ingested are far more dangerous than those that only increase exposure. This occurs through elements that enter food and water supplies. Such events have undoubtedly taken place throughout the Northern Hemisphere.

Another increase in the projected numbers of deaths would be the impact of radioactive fallout from Chernobyl on unborn fetuses. Children who are developing in the womb are particularly sensitive to radiation. There seems to be little that has been done to estimate the consequences of prenatal radiation exposure from Chernobyl.

These additional factors of ingested radioactivity and prenatal exposure, if known, would increase Gronlund's suggested figure of 26,000, but are unlikely to result in the estimated excess cancer deaths exceeding the upper confidence interval of 57,000.

Other Points of View

Others have adopted the approach taken by Gronlund in analyzing excess cancer deaths from Chernobyl. Most notable are leading expert on nuclear issues and National Medal of Science recipient Richard Garwin, as well as M. V. Ramana of the Woodrow Wilson School at Princeton University.

Linear No-threshold (LNT) Model

Scientists and others who have adopted this approach calculate the probability of excess cancer deaths due to radiation according to a linear relationship between exposure and impact. This linear relationship is assumed to apply to all levels of radiation—from very high to zero. This dose-to-impact model is known as the linear no-threshold (LNT) model.

The LNT approach has many critics. For example, some suggest that detailed scientific studies are not possible regarding exposure below 100 mSv, which is equal to 50 times the background radiation level; others have asserted that there is a radiation threshold value below which there is no damage done by radiation; and, yet others have claimed that at low levels of radiation, the effect is actually helpful rather than harmful. This last hypothesis is known as the hormesis model.

Up until 2006, the most definitive study of the health risks from exposure to low levels of ionizing radiation comes from the National Research Council. Known as BEIR VII–Phase 2, the study could not definitely conclude that the LNT was correct, but promoted the LNT model as the most plausible hypothesis when considering public health issues. However, BEIR VII–Phase 2 did not find support for either a threshold or for hormesis. During the 20 years since the publication of BEIR VII–Phase 2, there has been much discussion of these conclusions as well as important new research. Of particular interest is a review by the Nuclear Regulatory Commission (NRC), which has used LNT for some time as their guideline in evaluating radiation risks. The NRC considered changing from the LNT model to a hormesis model and invited public commentary on that issue.

In response, in October 2015, Director of the Radiation Protection Division of the US Environmental Protection Agency Jonathan D. Edwards wrote that "Biophysical calculations and experiments demonstrate that a single track of ionizing radiation passing through a cell produces complex damage sites in DNA, unique to radiation, the repair of which is error prone. Thus, no threshold for radiation induced mutations is expected and indeed none has been observed." Edwards went on to quote BEIR VII and noted that, since its publication in 2006, there have been several studies that reinforce its conclusions about the LNT model's correctness. Edwards endorsed the LNT model and explicitly rejected the possibility of hormesis.

While there are strong advocates for the hormesis model (Professor Edward Calabrese of the University of Massachusetts Amherst being the most prominent), there is little support in the wider scientific community for this hypothesis. It is significant that Edwards's response posits a physical model for the efficacy of a single track of ionizing radiation in damaging critical cells in a manner that could lead to the development of cancer.

Size of Radiation Levels

Arguments also exist that the effects of radiation levels below 100 mSv are too small to measure directly and therefore should be ignored. This notion is at odds with how other areas of science, most notably, general relativity, view the reality of small disturbances in nature. Such a counter-example emerged recently when Einstein's general theory of gravity was validated when gravitational waves were detected by an extraordinary optical apparatus known as the Laser Interferometer Gravitational-wave Observatory (LIGO).

The waves observed in February 2016 originated in the collision and merger of two black holes in deep space, which proved the existence of gravitational waves. However,

LIGO is only able to observe gravitational waves caused by the most massive and energetic of gravitational events in the universe, which, in addition to black holes, include events involving neutron stars. The fact that neither LIGO, nor any other apparatus, can observe small gravitational waves does not imply that the small gravitational waves do not exist. On the contrary; the observation of the large gravitational waves, together with the conceptual framework of Einstein's theory, solidifies belief in the existence of all gravitational waves.

The approach taken by Gronlund, Garwin, Ramana, and others has also been criticized on the grounds that the radiation levels to which they are imputing such deleterious effects are of the order of magnitude of background radiation. The notion that background radiation is itself benign is without foundation. Given that about 40% of the population acquires some form of cancer and that 20% of the population dies of cancer, it is plausible that background radiation contributes to acquisition of these cancers.

It is this high prevalence of cancer in public health figures that prevents observation of additional cancers due to specific sources, for example, low levels of radiation from cesium-137 fallout. However, when radiation becomes internal to the body such as through ingestion from the food supply, its effects are greatly magnified, and radioactive isotopes become manifest as dangers to health.

Background Radiation

It is worth providing perspective on enhanced danger in the case of background radiation. It is commonplace to consider background radiation as inconsequential, but radon is at least one manifestation of background radiation that has a significant impact on public health. Radon is a radioactive element that occurs in nature as an inert gas. It is produced by the radioactive decay of radium and is found in rocky soil containing shale, granite, uranium ore, schist, and limestone. It is quite common and can accumulate in basements. As a gas it can be inhaled and absorbed in the lungs where it undergoes radioactive decay in which an alpha particle, consisting of two protons and two neutrons, is emitted. These alpha particles damage cells and set the stage for lung cancer Radon causes between 15,000 and 20,000 deaths per year in the United States from lung cancer. It is the leading cause of lung cancer for nonsmokers. This death toll exceeds that caused by drunk driving (approx. 10,000 deaths per year in the United States). While background radiation may not cause obvious health problems due to irradiation to the body, background radiation is clearly not benign.

Radon is but one of the contributing sources in natural background radiation. Others include potassium-40 and carbon-14. Low levels of potassium-40 are found in bananas. Additionally, bombardment from outer space by cosmic rays is another source of natural background radiation.

Except for lung cancer from radon exposure, it is not possible to associate individual cancers with causal links to components of background radiation, but given the LNT relationship of radiation to cancer, these links await further investigation.

Nuclear Workers Research and the case for LNT

Page 290 of the Bier VII–Phase 2 report states that " . . . the most promising studies for the direct assessment of risk at low doses and low dose rates are those of nuclear workers who have been monitored for radiation exposure through the use of personal dosimeters." While there have been several such studies, a definitive report was published on Sept 9, 2015, by a team of researchers from the United States and Europe, with Professor David B. Richardson of the University of North Carolina as the lead author.

In this cohort study, 308,297 workers in the nuclear industry from France, the United Kingdom, and the United States detailed monitoring data for external exposure to ionizing radiation were linked to death certificate data. These data were acquired during the time periods of 1968–2004 in France, 1946–2001 in the United Kingdom and 1944–2005 in the United States. The report notes that "Follow-up encompassed 8.2-million-person years. Of 66,632 known deaths by the end of the follow-up 17,957 excess solid cancer deaths were attributed to radiation [. . .] Excess relative rate per Gy (dose) of radiation for mortality from cancer was estimated." The results of this study support an LNT relationship between radiation exposure and excess cancer, including low levels of radiation. This result answered one of the major objections to the Beir VII–Phase 2 study—that it posited that low-level doses over an extended period could be extrapolated from the high-level dose data from Hiroshima and Nagasaki. Questions were raised about the use of data from high-energy gamma rays from the atomic explosions and their short exposure duration being used to predict the health impact of low-energy radiation exposures over long time spans. This analysis of nuclear workers supports the generalizability of the Hiroshima and Nagasaki data. The study demonstrates that low doses over long time periods has the same impact as equivalent high doses over brief time periods.

Both the use of personal dosimeters to measure exposure as well as access to death records buttress the conclusion that the LNT theory is valid. The very large numbers of workers involved in this study provide conclusions with strong statistical validity. Note that the results of US, French, and UK research, as well as ten other studies, have tested the LNT theory during the 20 years since the publication of Bier VII–Phase. 2. While none of the other studies are as compelling as this, they all provide support for the LNT theory.

Conclusion

This chapter suggests that the Gronlund estimates, taken together with the recent studies of low-level radiation impact on nuclear workers, provide a compelling case for a conclusion that approximately 26,000 cancer deaths are attributable to the Chernobyl accident. This is more than six times the figure quoted by UN agencies and less than ten times that quoted by Greenpeace. These order of magnitude discrepancies cry out for further clarification.

More than 30 years have passed since the Chernobyl accident, which remains the world's most catastrophic nuclear reactor accident that the world has experienced. As such, it is a reference used to frame ongoing discussions about energy policy. Given that nuclear power could potentially provide a path away from global warming, the consequences of how Chernobyl is perceived are formidable. The number of deaths from accidents tends to short-circuit thinking and provide an oversimplified surrogate for detailed analysis.

We only need to look to Germany to find a major country that has eschewed the use of nuclear power. The Green Party influence was significant in bringing about that development, with their use of hundreds of thousands of deaths from Chernobyl included in their discourse. In contrast to such claims, we find that the UN agencies of the Chernobyl Forum continue to cite numbers in the range of 4,000 as the likely fatal casualties arising from the accident. Given that the most scientifically valid analysis is approximately ten times larger than that of the Forum and ten times smaller than that of Greenpeace, the public is placed in a quandary. Rational decision making in democratic societies deserves better.

When the Forum published its conclusions, the press spread that point of view as authoritative. On September 8, 2005, the *New York Times* published an editorial with the headline, "Chernobyl's Reduced Impact," in which they stated that the Forum had presented

> . . . the consensus of eight United Nations agencies, including those responsible for health, the environment and nuclear power [. . .] In the long run, the experts predict, some 4,000 emergency workers and residents of the most contaminated areas may die from radiation-induced cancer. That qualifies Chernobyl as a serious accident but not a catastrophe.

With wildly divergent views being expressed by groups that have made up their minds both for and against nuclear power, public interest is not served by having this 2005 editorial from the *New York Times* stand as a guidepost for a conceptual framework about Chernobyl.

It would better serve the public to have a group of independent scientists review the history of the past 30 years and the latest research on the impact of long-term low-level radiation exposure to come forth with an informed perspective on this subject. Such a review might be best done in a manner and by those with the expertise of the committee that was assembled by the National Research Council of the National Academies to prepare the Beir VII–Phase 2 report, which was an assessment of the health risks from exposure to low levels of ionizing radiation. That group consisted of 18 scientists from the United States, Germany, Canada, the United Kingdom, The Netherlands, and France. As mentioned, an analysis of deaths arising from Chernobyl would not be complete without representation from Russia, Ukraine, and Belarus.

Given that various international organizations can appear biased, such an undertaking would enjoy maximum credibility if it were done under the aegis of a foundation rather than a governmental organization.

Thirty years is also an opportune moment to assess the many other ways in which the Chernobyl disaster affected those regions in which radioactive fallout was deposited. Illnesses other than cancer were activated and large swathes of land were made uninhabitable. Significant psychological clouds also engulfed the region. The *New York Times* commented on the mental-health consequences of Chernobyl, declaring that "People from the region are anxious and fatalistic, based upon a greatly exaggerated view of the risks that they face. The result can be drug and alcohol abuse, unemployment, and an inability to function." While the history and consequences of Chernobyl are multidimensional and complex, the core number of likely deaths remains a focus of attention. Just as the number of 2,996 has become associated with the events of 9/11, the public still awaits an appropriate reference number for Chernobyl.

Further Reading

Zhores Medvedev, *The Legacy of Chernobyl*, W. W. Norton & Company, 1990.
Courageous Russian scientist Zhores Medvedev's authoritative account of the Chernobyl disaster. His account stands out amid a plethora of accounts of this accident that inaccurately report its history.
Zhores Medvedev, *Nuclear Disaster in the Urals*, trans. George Saunders, W. W. Norton & Company, 1979.
This is an account of a serious radiological waste explosion in 1957 kept secret by the Soviet Union until revealed by Zhores Medvedev.

13

The Fukushima Accident

Public sentiment in many states has turned against nuclear energy following the March 2011 accident at Japan's Fukushima Daiichi Nuclear Power Station. The Fukushima accident was, however, preventable.

— James M. Acton and Mark Hibbs[1]

Nuclear Power in Japan: Overview

Japan, when measured by Gross Domestic Product and manufacturing output, is one of the leading countries in the world. It ranks second and third by these metrics, a remarkable achievement given its location on islands that lack traditional energy reserves. To secure its vast energy needs, therefore, Japan looks beyond its shores. Sprawling ports and terminals receive among the largest imports of liquified natural gas, coal, and crude oil on Earth. These imports comprise around 90% of the country's energy supplies.

Before 2011, Japan produced significantly more (around 20%) of its own energy, and it did so in large part with nuclear reactors. Despite being the one country to have experienced the devastation possible with the violent release of energy from atomic bombs, it became one of the most prolific in embracing the peaceful uses of controlled nuclear fission.

Motivated by the oil shocks of the 1970s, Japan embarked on a buildout of nuclear reactors as a strategic priority that, by the late 2000s, resulted in 54 units generating one third of the country's electricity.

Whether executives at the Tokyo Electric Power Company, the operator of the Fukushima Daiichi nuclear power plant, could have foreseen the events of March 2011 has been a topic of debate and litigation since.

Though lacking in oil and coal, the geology beneath Japan is far from inactive. The region is situated at the intersection of four tectonic plates—the Eurasian/Chinese, the North American, the Philippine, and the Pacific. As they move, they generate energy

---

[1] James M. Acton and Mark Hibbs, "Why Fukushima Was Preventable," *The Carnegie Papers: Nuclear Policy*, March 6, 2012. Carnegie Endowment for International Peace. https://carnegieendowment.org/research/2012/03/why-fukushima-was-preventable?lang=en, accessed January 30, 2025.

Nuclear Energy. Edward A. Friedman, Oxford University Press. © Edward A. Friedman (2025). DOI: 10.1093/9780198925811.003.0013

that frequently releases shockwaves and tremors and, from time to time, does so quite destructively. Devastating earthquakes and tsunamis continually impact the entire country.

Records of earthquakes in Japan date back more than 1,600 years. Prior to the 2011 Tōhoku earthquake, the most devastating seismic event in Japan's recorded history occurred on September 1, 1923, when a quake measuring 7.9 on the Richter Scale struck the Kantō region, which includes the greater Tokyo area. Hundreds of thousands of homes crumbled from the shock, burned in the ensuing fires, or wasted away in a tsunami reaching 12 meters high. Taking more than an estimated 140,000 lives, the event is known as the Great Kantō Earthquake.

Tokyo is on a fault line that, on average, leads to a major earthquake every 70 years. Acute awareness of this vulnerability in the aftermath of 1923 led to reconstruction initiatives to design and build structures that could withstand earthquakes. These buildings were integrated into a development plan that included networks of roads and trains. Areas of refuge were established in parks and public buildings built to particularly high safety standards.

Thus, as an earthquake-prone country, Japan became focused on earthquake resilience in its buildings and infrastructure. Following the destruction of many of Japan's cities during the Second World War, these methods were employed as cities were rebuilt.

The Tōhoku Earthquake and Tsunami

No level of preparedness, however, could have mitigated the impact of the Tōhoku earthquake. On the freezing afternoon of March 11, 2011, a magnitude 9.0–9.1 earthquake occurred 45 miles off the eastern coast of Japan. The resulting tsunami reached as high as 40 meters, tripling the height of the Great Kanto Earthquake tsunami, and careened toward Japan at 435 miles per hour. Residents in the coastal city of Sendai had fewer than ten minutes of warning. The human toll, though far smaller than in the century earlier despite a much larger earthquake, was nevertheless massive.

Official figures in 2021 put the death count at 19,759, with several thousand more injured or still missing. Note that this death count is the result of the tsunami and is unrelated to deaths that were a consequence of the Fukushima nuclear accident. The nuclear accident, in fact, is not implicated in any radiation deaths but it did have fatal consequences that were a result of the evacuations and chaos that it precipitated.

As for infrastructure, 120,000 buildings were destroyed and a million more damaged. Airports, highways, bridges, and trains succumbed to the force, including the Tohoku Shinkansen high-speed rail, which was damaged at over a thousand locations. Power plants failed or shut down, as did water systems. Salt water flooded over 20,000 hectares of farmland, rendering it unfit for use for some time. Ports were submerged and destroyed, and around 29,000 fishing vessels were wrecked. An irrigation dam ruptured and washed away additional homes. Electronics factories, automotive manufacturers, and chemical plants were shuttered, and their inventories destroyed. In one of the higher-profile industrial accidents, the Cosmo Oil-owned Chiba I Refinery caught fire and raged for 10 days.

The earthquake triggered the automatic shutdown of 11 nuclear reactors in the region; ten of these weathered the events without incident or structural damage. At two sites, Fukushima Daiichi and Fukushima Daini, flood waters rushed over the top of the seawalls, flooding the backup generators needed to keep the reactors cool after the shutdown, as they continued to produce residual heat. At Daini, competency and professionalism by the site superintendent Naohiro Masuda and 400 employees spared the plant from a major accident.

In all, the nuclear safety record during the Tōhoku earthquake constituted an out-standing accomplishment. Blighting that record was the Fukushima Daiichi accident, which cast a shadow of fear over not just Japan but also much of the world. Though no one was killed directly in the accident or by the radiation it released, the event was a catastrophe of emergency management—one with real human cost—that shed light on the need for deep reforms and improvements in the Japanese nuclear sector.

Accident at Fukushima Daiichi

Fukushima Daiichi was the site of six of the 54 nuclear power reactors operating in Japan at the time of the accident. These were General Electric (GE) light-water boiling-water reactors (BWRs) constructed over a 12-year period between 1967 and 1979. Units 1 through 5 used the same basic design, while Unit 6 incorporated newer elements. Fate-fully, GE did not incorporate a reinforced concrete containment enclosure in any of its BWRs, as was used in other reactor designs. Instead, they used a lower-cost structure that had been severely criticized by several of their engineers. In 1976, three GE engineers resigned from their positions citing shortcomings in the design of these BWRs.

On the day of the earthquake, Units 1, 2, and 3 were in full operation. Unit 4 was shut down for refueling while Units 5 and 6 were in dormant states to allow for main-tenance. The active units shut down automatically when the earthquake occurred, and emergency cooling systems engaged to remove the heat generated by the continuing fission reactions.

Fifty minutes after the earthquake, however, the tsunami, carrying the debris it had accumulated during its journey inland, struck the reactor buildings. The wave at that point was 43–46 feet (13–14 meters) high, dwarfing the 18.7 foot (5.7-meter) high sea-wall that had been designed to protect the emergency generators. Thus, reactors that had managed to safely maneuver the worst earthquake in recorded Japanese history experienced catastrophic failure from a too-short seawall.

Water flooding into the Fukushima station caused a loss of all electrical power, includ-ing that from the emergency diesel generators. In a cascade of events, cooling pumps went out of commission, leading Units 1, 2, and 3 to overheat and ultimately melt down.

The interaction of the overheated water in the reactor core triggered the oxidation of the fuel element containers, releasing hydrogen gas—the same consequence that caused nationwide panic during the Three Mile Island nuclear incident. Unlike Three Mile

Island, however, the accumulation of hydrogen gas at Fukushima triggered explosions in each of the three hot reactors, destroying the enclosures and releasing radioactive material into the environment. While it cannot be said with certainty, perhaps if these reactors had been built with the reinforced concrete containment structures for which the GE engineers who resigned in 1972 had advocated, these explosions might have been avoided or mitigated.

Running from the Plume

The purpose of the ensuing evacuation was to avoid excessive exposure to the radioactive materials the explosions sent into the atmosphere. Plumes carrying fission products spread across the region and were observed internationally with highly sensitive equipment. Of concern were two prominent radioactive isotopes within the plumes: ^{131}I and cesium-137.

The risk of these isotopes derives from their ability to enter the human body through common biological pathways. Mineral iodine is essential to the healthy functioning of the thyroid. Through similar chemistry, radioactive ^{131}I can accumulate within the body, increasing the risk of thyroid cancer. Some consolation comes from the fact that thyroid cancer is highly treatable and not life-threatening. As many as 97% of cases, when left untreated, are non-fatal, with treatment leading to a 99.9% survival rate. Cesium-137 follows a similar biological pathway to potassium, dispersing uniformly through the body's soft tissue. With a half-life of eight days, ^{131}I disappears within several months of the accident. However, cesium-137, with a half-life of 30 years presents a more long-lasting hazard for individuals and for the environment.

Initially, at 9:23 p.m. on March 11 authorities ordered a 3-km radius evacuation involving 1,900 residents. As the situation deteriorated, the evacuation region was expanded, forcing a second relocation. The morning after the event, the evacuation zone was expanded to 10 km, and later that day, to 20 km, involving 78,000 residents. Eventually, around 150,000 people were evacuated because of the Fukushima accident.

There was mass confusion and miscommunications related to the evacuation, resulting from breakdowns in communications infrastructure, disrupted chains of command, and the fact that many municipalities issued their own evacuation orders independent of those from central authorities. Additionally, many people made personal decisions to leave the area, adding strain and congestion to critical infrastructure. This chaotic situation resulted in large numbers of people having to move multiple times, surviving only on the bare necessities they managed to grab from their homes. Of those residing within the nearest 2-km radius. 20% had to relocate more than six times. In some cases, families were unwittingly relocated to areas of higher radiation exposure and left without further communication for over a month. A prolonged "shelter in place" order within a 30-km radius zone stoked tension, fear, and paranoia.

Within the 20-km evacuation zone, there were about 2,220 patients and elderly people residing in hospitals and nursing homes. The trauma of moving the elderly and sick proved fatal for up to 44 people. Failures of infrastructure, including the absence of electricity. led to additional deaths, as did medical care interruptions that in some cases lasted for months.

The combined effects of the nuclear accident, the earthquake, and the tsunami were a heavy burden. A survey of a sample of evacuees revealed that 60% believed that their health and the health of their families had deteriorated after the evacuation. Perhaps as many as 1,600 deaths can be attributed to evacuation-related circumstances. Of those surveyed, 50% were found to be living apart from family members with whom they lived prior to the disaster. At least a third also reported a significant loss of income. This stress and disruption led to an increase in the number of suicides among evacuees, of which there have been 83 noted. This number is not surprising. While the suicide rate in Japan is about 18 suicides per 100,000 population, which is outside the top ten countries in the world, it quite high for a high-income country. Suicide is the leading cause of death for Japanese men between the ages of 20 and 44 and for women between the ages of 15 and 34.

Lasting Confusion

Though unclear in the confusion of the disaster, extensive, long-term monitoring of the radioactive fallout from the three reactors indicates that the overall impact to human health and the environment was low. The exposure levels did not threaten health, and no fatalities were observed that could be attributed to radiation. Such findings have been independently confirmed by the United Nations, the World Health Organization (WHO), and the International Atomic Energy Agency (IAEA).

Rather, inept evacuation planning, poor and misleading communication, and the ensuing chaos, confusion, and psychological distress escalated a severe but non-deadly industrial accident to a national catastrophe that claimed well over a thousand lives. It is probably the case that much of the evacuation was counterproductive.

The most harmful elements of the situation—miscommunication, misconceptions, and confusion leading to fear of the effects of radiation—propagated more of the same, with resounding effects across the world. The Fukushima nuclear disaster was, and continues to be, cited as a reason to fear nuclear power. In Germany, the event was a windfall for long-term advocates against nuclear power, who used it to successfully press for the adoption of a nuclear phaseout policy in the country.

More than a decade after the event, press reports continue to demonstrate confusion about the event, conflating the impacts of the tsunami and earthquake with those of the radiation release, or even entirely misconstruing the order in which the tsunami and nuclear accident occurred.

Was the Fukushima Disaster Preventable?

The accident kicked off more than a decade of investigation and litigation, resulting in several notable reports analyzing the accident and separate civil and criminal trials against the then-executives of the Tokyo Electric Power Company (TEPCO). A fundamental question in the trials and in the broader public discourse has been: was the disaster preventable?

If limited to whether the disaster was technically preventable, the answer is clearly yes. The safe navigation of the earthquake and tsunami by ten nuclear reactors in the region, including four reactors at Fukushima Daini (which experienced similar flooding to Daiichi), is a testament to the accident's preventability even with the technology of the time. Any number of added technical safeguards were available, for example: the construction of a taller seawall to prevent flooding in the first place; the placement of backup generators at a higher elevation, out of harm's way even in the case of flooding; or the use of reinforced concrete containment structures as a last line of defense.

Despite the minimal radiological harm to the public, TEPCO was far from blameless for the event. Investigation into why the accident was not prevented revealed inexcusable lapses of regulatory and corporate responsibility. Reports conclude that the Fukushima nuclear accident was not the result of unforeseeable events whose sheer force and complexity overwhelmed an otherwise diligent plant operator adhering to strict regulation. Rather, it was an event that could have, and should have, been prevented.

Published in 2012, the official report of the Fukushima Nuclear Accident Independent Investigation Commission, established by the National Diet of Japan, firmly contended that the events of March 2011 were a "manmade disaster." The Commission found that TEPCO failed in its responsibilities as a private corporation and operated with a severely inadequate approach to risk management. However, the problems that precipitated the disaster were above all systemic, involving "collusion between the government, the regulators and TEPCO, and the lack of governance by said parties."

The report argued that blaming individuals for the accident, rather than focusing on systemic reform, would amount to merely "cosmetic solutions." In a scathing denunciation of leadership across Japan's nuclear sector at the time, the report stated that "The underlying issue is the social structure that results in 'regulatory capture,' and the organizational, institutional, and legal framework that allows individuals to justify their own actions, hide them when inconvenient, and leave no records to avoid responsibility."

The civil and criminal trials that ensued, however, indeed focused on the culpability of specific executives at TEPCO. In the arguments made for their defense, the executives centered on the question of whether they could have foreseen a tsunami of such great height, and whether it was reasonable to expect them to have built the towering seawall necessary to guard against such a low-probability event.

The evidence uncovered to answer this question did not reflect well upon TEPCO. While the decision to build an 18.7-foot protective wall—less than half the height of the tsunami in March 2011—may have been reasonable at the time Fukushima began construction in 1976, there is ample reason to believe that the original design should have been judged inadequate prior to 2011.

A great deal had been learned about the dynamics of tsunamis and the history of tsunamis in the Fukushima region during that 35-year interval. A turning point in the global attention paid to tsunamis occurred December 26, 2004, when one of the largest tsunamis on record left a huge toll of death and destruction across 14 countries. In

Sumatra, Indonesia, waves as tall as 98 feet (30 meters) crashed ashore, killing over 150,000 in the country. Tens of thousands more died on the coast of Chennai in India, on the island nation of Sri Lanka, and in Thailand. Deaths were reported as far as South Africa. Virtually every country with shores on the Indian Ocean was affected.

The total death toll for the tsunami approached a quarter million. This gut-wrenching figure prompted a surge in the field of tsunami studies. Before March 11, 2011, enhanced research was pursued on the history of tsunamis in regions of concern, along with in-depth reviews of the modeling methods then used to predict them. These studies led to the development of revised guidelines for construction designs.

Two detailed studies of the Fukushima accident document these lapses that could have been avoided: in 2012 Acton and Hibbs published "Why Fukushima Was Preventable" was published, and 2015 saw publication of "The Fukushima Accident Was Preventable," which identified errors in the earlier 2012 article but largely agreed with its conclusions.[2]

TEPCO had commissioned in-house studies in 2008 and 2009 whose models showed potential local tsunamis ranging between 20 and 32 feet (6.1 and 10.2 meters) in height, exceeding that of the 18.7-foot (5.7-meter) high seawall. In each case, TEPCO either ignored the reports or left them in limbo for years, finally reporting the results of the 2009 study to the Nuclear Industrial Safety Agency (NISA) less than a week before the March 2011 tsunami.

Adding to the lapse of professionalism, Synolakis and Kanoglu suggest that these tsunami studies themselves underestimated the possible tsunami heights based on haphazard methodological errors. These methodological errors persisted in the absence of basic regulatory requirements for disaster modeling, which left it up to nuclear sites to craft their own assumptions about potential natural disasters. Synolakis and Kanoglu note that, "to any experienced scientist or civil engineer, it is inconceivable that there have been different earthquake designs for [nuclear power plants] at such close distance from each other."

In fact, these studies' warning of tsunamis higher than 18 feet (5.7 meters) were not even required by the regulator. A 2015 IAEA report pointed out that:

> The vulnerability of the Fukushima Daiichi Nuclear Power Plant to external hazards was not reassessed systematically and comprehensively during its lifetime. At the time of the accident, there were no regulatory requirements in Japan for such reassessments [. . .] The regulatory guidelines in Japan . . . were generic and brief and did not provide specific criteria or detailed guidance.

Given the history of tsunamis in Japan, the lax attitude toward engineering safety analysis is striking. Concluding that the Fukushima accident was preventable given the technology and information available at the time, Acton and Hibbs suggest that:

---

[2] Costas Synolakis and Utku Kanoglu, "The Fukushima Accident Was Preventable," *Philosophical Transactions* A373: 20140379 (2015).

In the final analysis, the Fukushima accident does not reveal a previously unknown fatal flaw associated with nuclear power. Rather, it underscores the importance of periodically reevaluating plant safety in light of dynamic external threats and of evolving best practices, as well as the need for an effective regulator to oversee this process.

Outcome of the Trials

To the courts, the culpability of TEPCO executives straddled the relative burdens of proof required in civil and criminal trials. While continuing to operate the plant in the presence of risks fell short of criminal negligence—an acquittal upheld in 2023 by the Tokyo High Court—the executives were nonetheless found guilty in civil court of a reckless lack of due care that caused an otherwise preventable disaster.

The civil lawsuit, brought by TEPCO shareholders, led to a decision on September 30, 2020, in which four executives were ordered to pay damages totaling 13.32 trillion yen, or US$97 billion (about $300 per person in the US). While it was recognized that these individuals could not possibly pay that amount, the victory for the plaintiffs was considered a major accomplishment.

The high court in Tokyo also considered the role of the national government's Nuclear and Industrial Safety Agency that had administrative oversight responsibility of TEPCO operations. The high court ruled that the Agency bore equal responsibility for the accident and that the accident was preventable. In its ruling the court asserted that both TEPCO and the Agency were more concerned about increasing the plant's operational expenses than they were with the issue of safety. The high court denounced the Agency for accepting the utility's insincere verdict on risk assessment and failing to fulfill its role as a regulatory body.

Escalation to the Level of Disaster

TEPCO, the regulator, and the government could have prevented the nuclear accident at Fukushima. Focusing too much on the failures at the site, however, risks understating the role of the sociological response, which included the evacuation of nearby residents and the unanswered or poorly answered questions about the effects of radiation. The result of this poor response was to escalate the situation from an industrial accident to a disaster.

The earthquake and tsunami took close to 20,000 lives. The nuclear accident-inspired fear and disorder took as many as 1,600 lives. Actual radiation exposure took none. Whether Fukushima could have remained a harmless industrial accident by either forgoing the evacuation or coordinating it better is a counterfactual scenario impossible to prove. Nor would a lack of public health impacts in such a scenario exonerate TEPCO and the Japanese nuclear regulator from dire shortcomings. Yet in analyzing Fukushima for lessons on the (mis)management of nuclear accidents, it is hard to avoid the role that overreaction, as the companion of under-preparedness, had in the "manmade disaster."

Impact on Japan's Energy Policy

After the Fukushima accident, Japan shuttered its entire nuclear fleet, totaling 54 reactors. Domestic electricity production dropped precipitously, with the lost nuclear power being replaced almost entirely with imports of liquified natural gas (LNG). The country initiated massive spending on new gas infrastructure, including the Soma LNG import terminal in the Fukushima prefecture. Security of natural gas supply became an even more dominant focus of Japanese energy policy (Figure 13.1). Furthermore, electricity-related carbon emissions rose substantially, peaking in 2013 (Figure 13.2).

In the time since, a reduction in the use of coal, the installation of solar panels, and the restart of some nuclear plants has brought emissions back to pre-2011 levels. However, energy security remains a major concern. When the European energy crisis began in the winter of 2021, worsened by the Russian invasion of Ukraine, tightening global supplies of natural gas sent Japan into electricity shortages.

Pragmatic energy security considerations drove the restart of two nuclear reactors in 2015, despite stark public opposition. In July 2022, amid an energy crisis, Prime Minister Fumio Kishida announced a full reversal of the country's nuclear phase-out policy. The government stated that the country would restart nine of its nuclear reactors by the

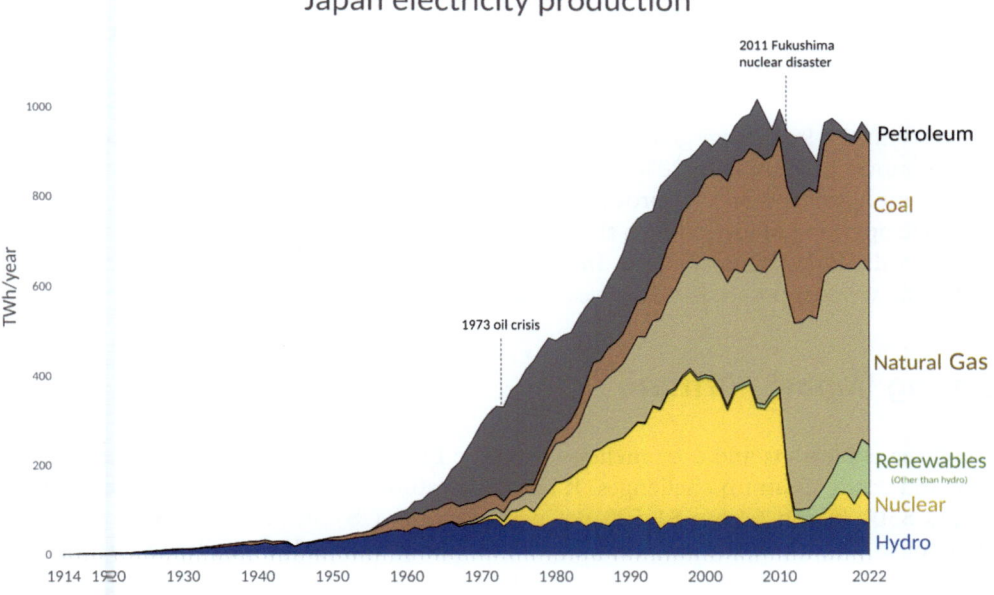

Figure 13.1 *Energy sources for Japan's electricity production (1914–2022).*

Reproduced from Kaj Tallungs (2021). Wikimedia Commons. https://commons.wikimedia.org/w/index.php?curid=102647716. under a Creative Commons Attribution-Share Alike 4.0 International (CC BY-SA 4.0).

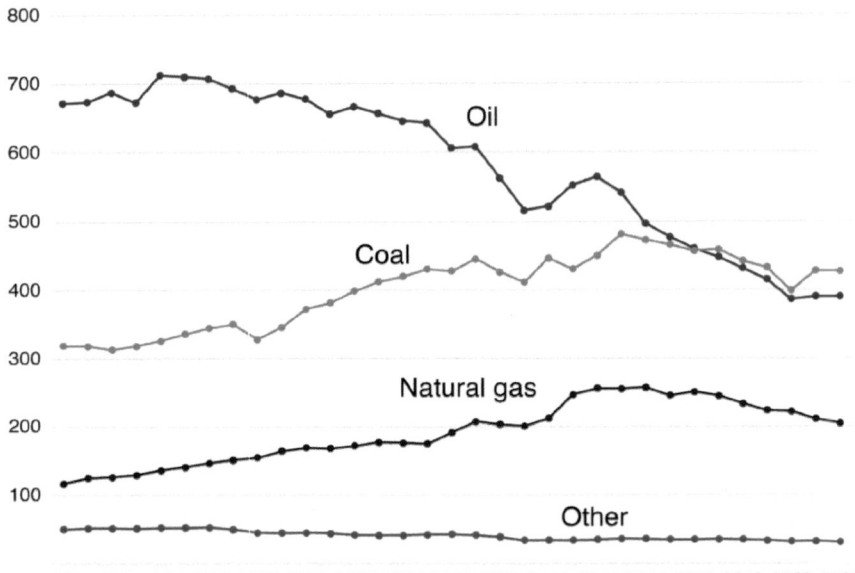

Figure 13.2 *Carbon dioxide output by fuel source or industry for Japan, 1990–2022 (in millions of tons).*

Source: Data from Our World in Data by Ritchie, H., Rosado, P., and Roser, M. (2023). Energy. https://ourworldindata.org/energy. under a Creative Commons Attribution 4.0 International (CC BY 4.0).

winter, with the long-term goal of returning the share of nuclear electricity generation in the country to more than 20%. As of late 2023, the country has restarted 11 reactors, with 16 more in the restart approval process.

Public opinion polling in Japan shows an easing of anti-nuclear sentiment, which was high after the Fukushima accident. In 2022, 47% favored the restart of nuclear reactors, while 30% were opposed. Just four years earlier, those numbers were reversed.

Lasting Impact—Tritium

The three explosions and core meltdowns at the Fukushima power plant created long-term and costly clean up challenges. It has been estimated that the cost of cleanup will exceed $500 billion (about $1,500 per person in the US) and take 30 to 40 years to accomplish. The three melted cores, while presenting formidable hurdles to restoration of a safe environment, are at least locally situated.

The melted fuel and radioactive debris require continuous cooling supplied by water pumped in for this purpose. Added to this water is groundwater that seeps into the site from the surrounding environment and rainwater that falls on the damaged site. Water from these sources is contaminated with radioactive elements.

Safe disposal of this water is one of the main problems being tackled by managers of the cleanup. From the time of the accident in 2011 until 2023, this radioactive water was stored in tanks at the reactor site. After accumulating hundreds of millions of gallons in over a thousand tanks, the site could no longer continue a storage strategy. Beginning in 2023, TEPCO began to release this water into the ocean after its treatment and dilution.

An advanced liquid processing system (ALPS) was developed for this purpose. ALPS is a pumping and filtration system that uses a series of chemical reactions to remove 62 radionuclides from contaminated water. However, tritium cannot be removed from contaminated water using ALPS because of its chemical similarity with the water.

Tritium is an isotope of hydrogen. While a hydrogen atom contains one proton and one electron, tritium contains one proton and two neutrons. Given that it is chemically identical to hydrogen it combines readily with oxygen to form a type of water. This isotope is unstable, and one of the neutrons spontaneously transforms into a proton by emitting an electron, known as a beta emission. The phenomenon takes place with a half-life of 12.33 years, so it remains present in the environment for a considerable time. In a year's time approximately 5.5% of a given quantity of tritium will decay. The decay product of tritium is an isotope known as helium-3.

The energy of the emitted electron is low and can travel only about a quarter of an inch in air. It is used in the manufacture of radio luminescent lights, watches, and self-illuminating key chains, all of which are safe as the beta emission is incapable of penetrating the outermost layer of human skin. If ingested, however, its radiation can damage internal organs. For that reason, it is not benign, and its presence in water and fish needs to be regulated.

The continued presence of tritium in the stored Fukushima water has been seized upon by critics such as Greenpeace, which claims that its release will have "serious, long-term consequences."

In high concentrations, tritium can indeed be harmful. However, the concentration of tritium in a liquid environment can be brought to a biologically safe level through dilution. That is precisely what is being done, with the result that the water being released shows a tritium level well below the existing standard for drinkable water.

The WHO recommends a limit of 10,000 Becquerels per liter (Bq/l) of tritium for drinking water, where a Becquerel is a unit of radioactivity defined as one nuclear decay per second. At the request of the Japanese government, TEPCO will dilute the water held in its storage tanks by a factor of 100 before release, achieving a tritium concentration of 1,500 Bq/l, 1/7 that of the WHO limit for drinking water and 1/40 that of the regulatory limit for nuclear plant discharges.

The rate of discharge is planned not to exceed 22 TBq (terra Becquerel or a million-million Becquerel) per year, which was the target for the Fukushima station during normal operation. This is a normal rate for gigawatt-scale nuclear plants, as all water-cooled nuclear plants produce tritium during operation in proportion to the plant's energy output. Extensive monitoring around the world has found no impact on human health or the environment related to these ordinary discharges of tritium.

Even when the release of tritium is orders of magnitude higher, the environmental impact is negligible. Research on tritium concentrations in the English Channel near the French nuclear fuel reprocessing site at La Hague, which releases around 10,000 TBq per year (450 times the rate at which Fukushima water will be released), found a maximum observable concentration of tritium of about 30 Bq/l, still only 0.3% the WHO recommended limit for drinking water.

Despite the lack of a scientific basis for claims that the release of tritium will pose environmental or health threats, South Korea and China have been particularly harsh in their criticism of Japan for their release of tritium into the Pacific. After the release of the water began, China banned all seafood imports from Japan, prompting the United States military to buy Japanese seafood in bulk to offset the economic consequences. South Korea has maintained its seafood import ban from various locations in Japan enacted after the initial Fukushima accident. This is a bizarre action from China, given the fact that both South Korea and China, as well as Russia, the United States, Canada, the United Kingdom, and France all have reactors regularly emitting tritium into nearby bodies of water at far greater rates than will result from Fukushima.

The greatest threat facing those who rely on the waters off the coast of Japan for the fish that provide their livelihood is unsafe exposure to tritium. Misinformation and anti-scientific responses have resulted in an unnecessary loss of business.

In response to the intense scrutiny being addressed to the tritium release into the Pacific, TEPCO has engaged the IAEA as ongoing monitors of the program. As of late 2023, independent sampling by the IAEA determined that the levels of tritium off the coast of Japan were consistent with the measurements published by TEPCO and that these levels were within operational limits.

Further Reading

David Lochbaum, Edwin Lyman, Susan Q. Stranaham, and the Union of Concerned Scientists, *Fukushima: The Story of a Nuclear Disaster*, New Press, 2014.
A detailed review of the events that took place at Fukushima with an account of the aftermath. Written by analysts associated with the Union of Concerned Scientists, which has a record of preparing critical assessments of nuclear energy initiatives.
James Mahaffey, *Atomic Accidents: A History of Nuclear Meltdowns and Disasters from the Ozark Mountains to Fukushima*, Pegasus Books, 2014.
James Mahaffey, a long-time advocate of nuclear energy, examines the causes and lessons to be learned from nuclear accidents that took place from the beginning of the use of this technology to the events at Fukushima.

14

Solar, Wind, and Battery Power

Even the darkest night will end and the sun will rise.

—Victor Hugo, *Les Misérables*[1]

Quest to Eliminate Carbon Emissions

The task of replacing fossil fuels, necessary to eliminate carbon emissions from their combustion, is truly daunting. As of 2023, fossil fuels supplied a staggering 81% of global energy consumption, with oil (33%), coal (27%), and gas (21%) dominating the energy mix. The remaining 19% comes from solar and wind (11%), nuclear (6%), and hydropower (2%).

To eliminate oil, coal, and gas as the primary carbon-emitting fuels providing the world's energy, some combination of increased solar, wind, and nuclear energy generation is needed. While other mechanisms for carbon elimination (e.g., carbon capture and use of biomass fuel) will contribute, solar, wind, and nuclear power will be the dominant contributors. Meeting this challenge is a monumental undertaking.

This chapter provides perspective on the viability of wind and solar power to meet this challenge. In recent years there has been a surge of new installations, which leads us to ask to what extent this increase might be sustained and what the balance might be among the three main groups of electricity generators at mid-century.

In starting with solar power, note that energy from the sun can be harnessed directly as heat or as electricity with the use of solar photovoltaic cells, which make electric current from absorbed photons.

Generating electrical energy using direct sunlight with mirrors dates to the later part of the nineteenth century, when French mathematics teacher Augustin Mouchot was concerned that the world supplies of coal would run out. He constructed a solar steam generator that powered an engine, which he exhibited at the World's Fair in Paris in 1878.

---

[1] Victor Hugo, *Les Misérables*, trans. Isabel F. Hapgood., Thomas Y. Crowell & Co., 1887.

Nuclear Energy. Edward A. Friedman, Oxford University Press. © Edward A. Friedman (2025). DOI: 10.1093/9780198925811.003.0014

Photovoltaic Energy

In 1839, Becquerel's discovery of the photovoltaic effect, which converts solar irradiation into electricity, brought on the birth of the principles of solar energy. The general definition of the photovoltaic effect is the formation of a voltage between two electrodes, separated by a solid or a liquid, upon illumination of the system via irradiant light. The first observation in 1876 that light can generate an electric current in selenium, and the first prototypical selenium/gold photovoltaic system with less than 1% conversion efficiency marks the birth of this revolutionary new technology in 1883, although the term solar cell was not coined yet. In 1904, 65 years after the photovoltaic effect was discovered, copper and copper oxide were incorporated as semiconductor junctions—the founding principles of modern solar cells. The brief pathway through the history of the discovery of the photovoltaic effect to the use of semiconductor junctions to create a solar energy conversion device can now help concisely define the term.

A single photovoltaic cell typically produces 1–2 watts of electrical power. The cells are connected into an array that forms a panel that can vary in size and efficiency and its output varies with the nature of the mounting, the location of the panel, and the weather conditions. For this discussion, we use an average output for a typical 17-square foot (1.6 square meter) panel of 300 watts per hour.

Assuming an average efficiency of 18% for a panel, representing the percentage of solar radiation energy that the panel can translate into usable electrical energy, 5,556 panels would be needed to generate 1 megawatt (MW) of electrical energy. This number of panels would occupy 8,800 square meters (about the area of a Manhattan city block). A thousand megawatts would then require 8,800,000 square meters (about the area of the Philadelphia International Airport) or four square miles. Actual projects show that land use can be, at times, much higher than this rough estimate.

The Bhadla Solar Park in the Thar Desert of Rajasthan, India covers an area of 56 square kilometers (about twice the area of Chicago O'Hare International Airport) and has an installed energy capacity of 2,245 MW of electrical power. It occupies 21.6 square miles.

The Karapinar Solar Power Plant in central Türkiye generates 1,300 MW of installed power and covers an area of 7.7 square miles.

One of the world's largest single-site solar plants, Noor Abu Dhabi, covers an area of eight square kilometers, or three square miles, with 3.2 million solar panels producing approximately 1 gigawatt (GW) of electric power at peak sunshine. This is notably high energy density for a solar PV project, owing to its ideal location in flat, sunny desert conditions.

In 2015 the world's largest photovoltaic power plant was Solar Star near Rosamond, California that used 1.7 million panels spread over five square miles (13 square kilometers) to produce 579 MW (AC) of electrical power.

In 2024 the largest installation producing solar voltaic energy was the Gonghe Talatan Solar Park in China's Tibetan Autonomous Region nearly 1.85 miles (3,000 meters) above sea level. This location receives intense sunlight for an average of 1,600 hours per year. The plant produces an annual power generation of 9.6 billion kilowatts (KW) from a total area of ten square miles (609.6 square kilometers).

We see that the amount of land occupied by a solar plant can vary a good deal depending on the location and the nature of the landscape. Solar panels can vary widely in terms of efficiency and mounting. In comparison, a nuclear plant producing 1,000 MW of electricity requires around one square mile of land (Figure 14.1).

As of 2022 there were 72 solar parks worldwide with just 15 having an output of 1000 MW or more. Of these, six were in China, four in India, three in the United Arab Emirates, one in Egypt, and one in Turkey. Thus, we see that there are no large solar parks in North or South America, Europe, or Australia.

Photovoltaic solar was responsible for 4.5% of the world's electrical energy in 2022, having grown 25% above its 2021 share of 3.6%. In this time frame, the world's photovoltaic electrical energy originates about equally from solar farms and distributed solar panels on rooftops and other buildings in urban locations.

The International Energy Agency predicts similar rates of increase in the use of photovoltaic electrical energy in the years going forward due to two major factors. One is the

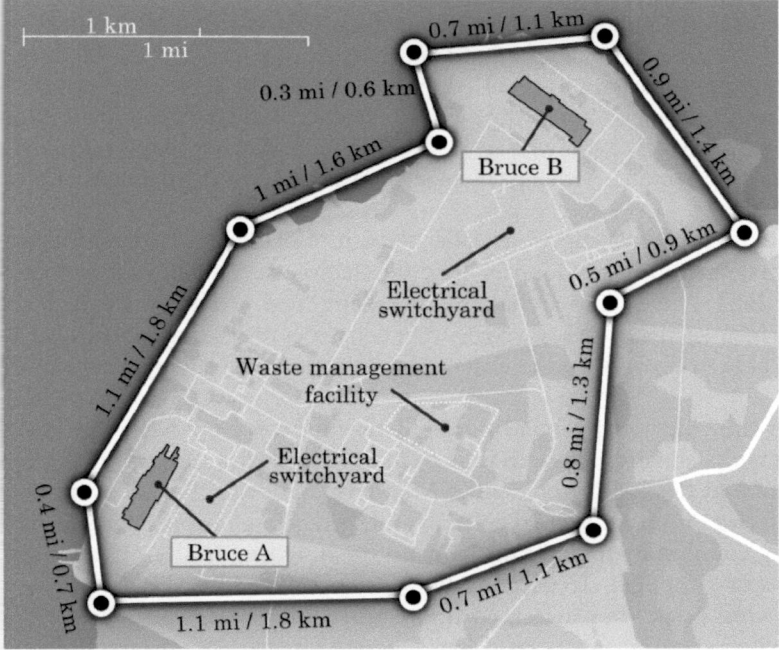

Figure 14.1 *Bruce Power Generating Station in Ontario, Canada, with four CANDU reactors providing 2832 MW of thermal energy.*

Reproduced courtesy of Alex Wellerstein (Stevens Institute of Technology, Hoboken, NJ, USA). Source: Map data from OpenStreetMap. (2024). https://osmfoundation.org/ under an Open Data Commons Open Database License (ODbL) v1.1.

high cost of fossil fuels and the declining cost of photovoltaic electricity. The second is the Russian–Ukrainian war, which disrupted supply chains for fossil fuels causing fossil fuel importers to increasingly value the energy security benefits of renewable energy.

The outlook is not as bright in developing countries where weak grid infrastructure and a lack of affordable financing hamper faster commissioning of multiple projects.

Note that China, on its own, is forecast to install almost half of the new global renewable capacity over the period from 2022–2027. The European Union (the second largest growth market after China) has been particularly affected by the war in Ukraine and plans implemented since the outbreak of the Ukrainian war call for the elimination of Russian fossil fuel imports prior to 2030. In the United States, the Inflation Reduction Act passed in August 2022 extended tax credits for renewables, including photovoltaic electric power through 2032. During this time frame there are also significantly expanded programs for acquisition of photovoltaic electric power in Brazil and India.

Wind Energy

The capacity of windmills to produce electricity is a consequence of Faraday's law of induction. Born in England in 1791, Michael Faraday was one of the most productive scientists of the nineteenth century and contributed to chemistry as well as physics. He received only the most basic school education and at 14, he became apprenticed to a bookseller and read many books during his seven years of service. While he had limited mathematical knowledge, he was an avid experimentalist and explored both chemical and physical phenomena, with a particular fascination for the properties of magnets and electrical currents.

His breakthrough investigations took place in 1831 when he found that when he moved a magnet through a loop of wire an electric current flowed in that wire. His demonstrations showed that a changing magnetic field produced an electric field. He used the principles that he had discovered to construct the electric dynamo. This device, which was able to convert mechanical energy into electrical energy, was a revolutionary invention.

It was in 1887 that the first wind-powered machine was employed to produce electricity, with systems constructed both in the United States and Great Britain. James Blyth built a system at his holiday cottage in Marykirk, Scotland. He used the electricity to charge batteries for his household lighting and then offered the surplus electricity to the neighboring community for lighting the main street. However, the villagers turned down the offer, as they thought electricity to be "the work of the Devil".

Use of wind power has increased dramatically in recent years due to it becoming economically competitive with fossil fuel generated power. Wind power technology has become more efficient. In 2022, wind supplied over 7% of the world's electricity. Wind power is most often produced at wind farms that consist of several hundred individual wind turbines. To minimize the environmental impact of wind farms on land, offshore placement is favored. However, the cost of offshore construction is significantly higher than land-based placement, with escalating costs and supply chain issues leading to the cancellation of several planned offshore wind projects on the east coast of the United States in 2023.

An individual wind turbine typically has blades that are about 165 feet (50 meters) in length and produce 2.5–3 MW of electricity. However, General Electric (GE), Vestas, and other wind turbine manufacturers have scaled up the size of wind turbines to the 6-MW and larger range, with blades exceeding 246 feet (75 meters) in length. Wind turbines operate when the wind velocity is greater than 6–9 mph and shut off for safety when the wind velocity exceeds 55 mph. Overall, they operate at 30–40% efficiency.

A wind farm must provide significant space between turbines for a given turbine to have wind currents that do not limit flow from neighbors. Overall, wind farms require up to 350 acres for a 1000-MW production facility.

Of the five largest wind farms in the world, four are in China, with the largest being the Gansu Wind Farm, which produces 20,000 MW of electrical energy. The next four are in the range from 1,600 MW to 3,000 MW.

The largest wind farm in the United States by installed generating capacity is the Alta Wind Energy Center in California's Tehachapi Mountains. Six hundred wind turbines generate a maximum output of 1,550 MW and occupy an area of land of 50 square miles. This is a land usage of 32 square miles per GW.

Energy Storage

While both solar energy and wind energy have attractive features, they share the common challenge of maintaining a steady supply of energy. Renewable energy sources allow us to capture energy flows in the environment, but apart from locking a wind turbine or otherwise preventing the generation of electricity from a renewable energy source, we cannot decide when they do so. This is a major challenge, as operating the sensitive system of the electric grid requires precise, instantaneous matching of energy supply with energy demand. Having either too much or too little electricity can disrupt energy markets or force grid operators to engage drastic procedures to maintain grid balance, for example, load shedding or blackouts.

The production of wind and solar energy is not a binary on–off process but rather varies with the intensity of the sunshine and wind. For this reason, they are called variable energy sources. A strategy for dealing with this variability is to store excess energy when the demand is low and tap into this stored energy when it is needed. There are multiple technologies that are available for this energy averaging approach.

Historically, the most widely used system for energy storage has been pumped hydropower, in which water is pumped from a lower reservoir to a location that is elevated. From there, the energy spent pumping the water uphill can be reclaimed later by reversing the process: releasing the water which, gaining speed from gravity, collides with a turbine to spin an electromagnetic generator to produce electricity. The process loses energy due to inefficiencies. Therefore, to remain economic, the facility must pump water when the electricity required to do so is cheap and release water to sell electricity when the price is high.

In 2023 the world total for pumped hydropower is about 168 GW, with about 25 GW in the United States. While pumped storage accounts for 96% of all grid-based energy storage in the United States the increase in renewable energy production is stimulating expansion of stored energy facilities. While pumped hydropower could possibly double in the United States, other mechanisms are likely to become dominant due to having fewer geographic limitations.

A mechanism that is widely used to store solar energy is based upon the use of molten salts. The Gemasolar Thermosolar Plant in the province of Seville, Spain was able to provide electrical energy from solar input for 24 hours per day for 36 consecutive days, thus proving the capability of molten salt technology to even out energy supply over periods of limited to no sunlight.

The most rapidly increasing energy storage systems use rechargeable batteries with lithium-ion technology. An example is the Aliso Canyon SCE Mira Loma Substation in Ontario, California, where there are stacks of 396 refrigerator-sized units providing 10 MW in each of two modules. This configuration can run for four hours. Other locations are in place using similar technologies using lead–acid, nickel–cadmium, and other combinations.

Nickel–hydrogen batteries provide 40 MWs of electricity for ten hours in Puerto Rico to support its industrial sector. Unlike a lithium–ion alternative, this nickel–hydrogen battery can operate in conditions of extreme heat and cold.

An unusual battery design is that of the so-called flow battery, which does not use solid electrodes but rather employs two liquids (one positively charged, and one negatively charged) separated by a porous membrane. The most advanced flow battery design are solutions of vanadium metal for both the positive and negative components, which can be accomplished due to the complex nature of vanadium ions. Flow batteries are particularly suitable for use in balancing fluctuations in wind farms. This technology is in use at the Sorne Hill Wind Farm in Ireland and in Tomari Winded w Hills Japan. Some advantages of Vandox flow batteries include absence of a limit on energy capacity; safe, non-flammable aqueous electrolyte; and ability to add modules as needed. Disadvantages include high cost of vanadium minerals; low efficiency compared with lithium–ion design; and the need for pumps to maintain flow of electrolyte solution.

With the prevalence of variable wind and solar generation, and the vast planned capacities, the demand for tools to store and dispatch energy has led to significant innovation. During the past 15 years the cost of batteries has fallen by more than 90%. While this is an impressive development, the commercialization of this technology remains a challenge.

Promise and Challenges

The use of solar and wind technologies is highly dependent upon access to viable grid connections. Places like Ontario and Alberta in Canada have pursued significant renewable agendas that were then halted due to rising prices and grid challenges. These are challenges to be solved. Balancing production with wind energy and the use of storage is one option, but these together still have trouble meeting demand alone. US Department of Energy reports show that the presence of zero-carbon baseload nuclear generation in net-zero energy scenarios dramatically reduces the magnitude of wind, solar, and batteries required to achieve grid operation, with a several-fold reduction in overall costs, land use, and materials consumption. Governments around the world have likewise recognized the significant role that dispatchable low-carbon generators will play. While solar, wind, and batteries are being built and will play a significant role in the adoption of low-carbon energy, theories that these technologies alone are sufficient to power modern civilization are fringe and have been widely discredited.

In closing this chapter, some sobering observations are shared from *Sustainable Energy: Without the Hot Air* by British mathematician and physicist Sir David MacKay, in which he analyzes the ability of various energy sources to meet societal needs, with simple calculations. He includes an evaluation of the theoretical maximum of inland wind energy that can be delivered on a per person basis in the UK (20 KW), compared with the average amount of electrical energy per day that is used by an individual in a day (250 KW). He goes on to show that the maximum amount of energy that could be obtained by using all the available coastal areas for additional wind turbines is only another 20 KW per day. His conclusion that all sources of renewable energy fall short of meeting the current energy usage makes a strong case for the inclusion of nuclear energy into the mix needed to achieve net zero.

Further Reading

Bruce Usher, *Renewable Energy*, Columbia University Press, 2019.
A primer on the evolving transition from fossil fuels to renewable sources of energy with an emphasis on wind and solar.
E. Calvin Beisner and David R. Legates, eds., *Climate and Energy: The Case for Realism*, Regnery Publishing, 2024.
A collection of essays that cast critical eyes on current initiatives aimed at ameliorating harm from climate change.

15

Generation-III Reactors

The atomic bomb is the second coming in wrath.

— Winston Churchill, 1945[1]

Overview

The shock of Chernobyl in 1986 brought nuclear reactor development to a standstill throughout the world. In Russia, it forced a rethinking of reactor technology, which created a hiatus of a decade before active production resumed. In the West, the Chernobyl shock deeply affected public opinion and caused a pause lasting several decades. Adding to the impact of Chernobyl was the explosion of three reactors at Fukushima in 2011 that reinforced antipathy to nuclear energy.

In Russia, it was seen that new designs were needed since the RBMK model of Chernobyl was no longer a viable option. Work began soon after the Chernobyl accident to develop new approaches to nuclear energy generation for Russian industry. Renewed activity in the west evolved more slowly.

After Chernobyl it was obvious to nuclear engineers everywhere that passive safety systems were needed. The goal for passive safety is to have a reactor respond to an accident without any action by operators or activation of machinery such as pumps, cooling devices, or other mechanisms that require electrical input. A true passive safety system would depend only on the forces of nature (e.g., gravity) that would open a valve or bring water flowing down to cool an overheated component. Differences in temperature could also maintain the flow of a liquid. Additionally, overheating could cause the melting of a sealant allowing a liquid to drain from a region in the reactor where it was supporting energy development to a location where it was no longer supporting the fission process.

This chapter reviews the development and use of reactors known as Generation III. Since this category does not have a universally recognized definition, we need to make clear that we are designating reactors as Generation III if they use water as a coolant and employ passive cooling. Included in this category are reactors that use both active and

---

[1] Taken from the Diaries of Lord Moran: "The Struggle for Survival 1940-1965" Lord Moran was the personal physician to Churchill.

Nuclear Energy. Edward A. Friedman, Oxford University Press. © Edward A. Friedman (2025). DOI: 10.1093/9780198925811.003.0015

passive cooling. Those reactors that employ passive cooling and do not use water as a coolant are discussed as Generation IV reactors.

The withdrawal of support for nuclear energy following the accidents was dramatic:

- Belgium built seven reactors in the 1970s–1980s, then none until two were started in 2023
- Canada built 24 from the 1960s–1980s, then paused until 2023
- The United Kingdom built 28 from the 1960s–1980s, then none until 2018
- The United States built 128 from the 1960s–1980s, followed by a hiatus until two began construction in 2013 which were not completed until 2023–2024.

In the United States, 41 reactors have closed during this period and there are ongoing discussions about the viability of several of the reactors that have been operating for 40 or more years.

There appears to be a pattern of retreat from nuclear energy in France, which has been a very pro-nuclear country and where 70% of its electrical energy has been produced by nuclear reactors. Government policy shifted in 2014 with a move to reduce the nuclear percentage to 50%. During the 1960s–1980s, 67 nuclear reactors were built in France and then none until a single new plant began construction in 2007.

The most dramatic development in the West has been the German policy of moving away from nuclear energy entirely. In 2002 the Green Party in Germany succeeded in having legislation passed to halt any new nuclear energy development and to close all existing plants. From the 1960s through to the 1980s, Germany acquired 31 nuclear reactors, mostly from Russia and by 2023 all but three plants have been closed. In 2022, Germany began to reexamine its anti-nuclear policy and postponed the closing of those three remaining plants,

Developments in China

Nuclear power emerged in China along with its other areas of industrial development that began in the 1980s. There were no nuclear reactors operating in China in the 1960s or 1970s. Construction for the first reactors in China started in 1985 with three reactors becoming operational in 1994. After Chernobyl there was a 12-year pause in nuclear construction that ended with the initiation of five new plants in the late 1990s. Since then, China has pursued an aggressive program of nuclear construction. From 2000, China has seen 51 nuclear reactors become operational and has started construction on an additional 22. This far exceeds the construction activity of any other country in the twenty-first century. Given the late start in construction of nuclear reactors, it is unsurprising that China has never shut down a reactor. In addition to the need for more electrical power to support a growing economy, China suffers from severe air pollution that must be reversed.

From around 2008, China has been pursuing implementation of Generation-III nuclear technology. Toward that end it engaged with Westinghouse to build four

Westinghouse Generation-III reactors, known as the AP1000, which is discussed at length as part of the US program. China began construction of the AP1000s in 2009. These reactors became operational in 2018.

With consulting input from Westinghouse, China has designed the CAP1400 (or the Guohe One), which is an expanded version of the AP1000 pressurized-water reactor (PWR). China intends to employ the CAP1400 in large numbers across the country, holds full intellectual property rights to this reactor, and plans to market it for export. Many components of the CAP1400 have been improved over prior reactor designs, including pumps, valves, and steam generators. The CAP1400 passed the International Atomic Energy Agency's (IAEA) Generic Reactor Safety Review in 2016, which will facilitate receiving approvals in various export destination countries. Construction of two CAP1400 reactors began in Shidaowan Province in 2019 and 2020 with completion anticipated by 2025. China anticipates mass production of the CAP1400 for domestic and export deployment.

Developments in Russia

In Russia, there were 39 nuclear reactors built from the 1960s to the 1980s. Following Chernobyl there were none built in the 1990s, which instead was a decade that saw re-evaluation of the nuclear designs both in Russia and in the West. However, given that Russian policy regarding nuclear energy was not influenced by public opinion in the same manner as in the West, Russia resumed production of nuclear reactors earlier than Western countries. Construction on one reactor began in 2008, followed by two in 2009, and a third in 2010. However, Russia did not wait until 2008 to resume nuclear reactor production for its export market.

Russia has surged ahead since 2000 in the export market of nuclear energy with an aggressive policy of providing full turnkey operations to countries with no nuclear experience or infrastructure. They also offer financing and nuclear waste treatment. The Russian nuclear corporation Rosatom has full state backing and is part of national geopolitical initiatives, particularly in Global South countries that nominally hold neutral positions between the West and Russia.

India falls into this category and in 2001 construction of a Russian reactor began in Kudankulam and was completed in 2013. This was despite active opposition from many in the local population who thought that the reactor would contaminate their fishing waters.

The Kudankulam reactor is distinctive because it is perhaps the first Generation-III reactor to become operational. Russia has had a policy of promoting a standard design that was first introduced in 1964, known as the VVER, a PWR for which the 1964 design underwent multiple improvements. Since basic design features were retained, there has been efficiency in its fabrication that has not been enjoyed by nuclear reactor manufacturers in the West. For example, the metal enclosure for the reactor core is fabricated in a factory and shipped by rail to construction sites or ports.

The Kudankulam reactor was built with a combination of active and passive safety features. The first stage of the passive operation consisted of water tanks that could flood the reactor vessel to provide cooling. The second stage involves secondary water tanks in addition to automatic condensation of the steam being generated by the reactor to flow back to the core to provide additional cooling. This process could operate for up to 280 hours (about 1.5 weeks). Also, water used to cool the contents in the spent fuel storage pool is available for diversion to cool the core for an additional 72 hours.

These passive safety features were gradually introduced into the VVER design. Through standardization, Russia has been able to assure prospective clients of reliability due to the excellent VVER reactor safety record. Also, Russia has standardized the design of the fuel elements and reprocessing for these exported reactors. A standard design also facilitates the development of effective programs for training operators.

In 2023, Russia was engaged in more than three dozen initiatives to build or negotiate construction of export nuclear reactors. This outreach is conducted by Rosatom, which became a state agency in 2007. It has responsibility for the country's nuclear power industry, nuclear weapons division, nuclear powered icebreaker fleet, and nuclear research institutions, as well as ensuring nuclear and radiation safety. The President of Russia sets Rosatom's strategic objectives and appoints its director and the members of its supervisory board. Rosatom manages more than 300 companies and organizations involved in all stages of its nuclear weapon and power production chain. This includes front-end nuclear fuel cycle activities (e.g., uranium mining, enrichment, and fuel fabrication); activities related to the construction, operations, and decommissioning of nuclear power reactors; and back-end cycle activities (e.g., spent fuel reprocessing and radioactive waste management).

Rosatom is also a major exporter of nuclear fuel. Its subsidiaries supply fuel to over 70 power reactors around the world and several research reactors. While most fuel contracts are with countries that operate VVER reactors,[2] Rosatom also exports enriched uranium to the United States, the United Kingdom, Belgium, France, Japan, and South Korea. This international outreach for fuel (as well as nuclear reactors) far exceeds the operations of any other country. Significant Rosatom activities are found in Europe, Asia, Africa, and South America. Since many of the nuclear relationships involve Russian financing, Russia can exert pressure on policies and political actions of countries worldwide.

Developments in the United States

In the United States, the Chernobyl accident halted construction of new reactors for civilian power. The two leading US reactor developers, General Electric (GE)–Hitachi and Westinghouse, pursued similar paths in creating original new designs for a

---

[2] Currently Russian VVER reactors are operating or under construction in Armenia, Bangladesh, Belarus, Bulgaria, China, Cuba, Czech Republic, Egypt, Finland, Germany, Hungary, India, Indonesia, Iran, Poland, Slovakia, Turkey, Ukraine, Uzbekistan, and Vietnam. In addition to these established relationships, Rosatom is engaged in nuclear-related discussions with Algeria, Bolivia, Cambodia, Ghana, Nigeria, Paraguay, Saudi Arabia, Sudan, Tajikistan, Tunisia, the United Arab Emirates, and Zambia.

Generation-III nuclear reactor with initial construction in Asia. GE–Hitachi built a Generation-III boiling-water reactor (BWR) in Japan that became operational in 1996, while Westinghouse built a Generation-III PWR in China that became operational in 2009. Westinghouse had an intermediate-sized Generation-III reactor in the 1990s that did not find a client for construction. The two leaders then followed different paths with GE–Hitachi engaging in the development of a small modular reactor, which Westinghouse did not pursue.

Around 1989–1990, Westinghouse recognized a need for a new safer generation of nuclear reactors. They put in motion a program to develop a reactor that incorporated innovative passive safety systems and proven technologies: the Advanced Passive 600-megawatt Electric reactor (AP600). This was the first Generation-III reactor in the United States.

The AP600 Nuclear Power Plant design by Westinghouse was under the sponsorship of the United States Department of Energy and the Electric Power Research Institute. The design team included several US and foreign companies and organizations. These included Bechtel, Burns & Roe, Southern Electric International, Oregon State University, UTE (Spain), ENRA Energy Research Center of Italy, and Initec of Spain. Through international cooperation involving entities that had expertise in every aspect of power plant operations, a comprehensive design was achieved that dealt with all components of a power plant. It not only had passive safety but also a vastly simplified design that facilitated construction and implementation. The plan for construction of the AP600 was to go from first concrete to fuel load in 36 months. Interactions with the Nuclear Regulatory Commission (NRC) began in 1992, which led to NRC design certification in 1999. Given the prevalent anti-nuclear environment, it is unsurprising that there were no orders for the AP600! One might say that this initiative by Westinghouse was a courageous act pursued in the interests of the company, the nation, and Western Europe.

In response to the failure of the AP600 to gain traction in the marketplace, Westinghouse decided to double down and produce a larger cost-effective reactor, the AP1000, having an output of 1000 megawatts (MW) electric. The NRC approved the AP1000 design in December 2005. In 2008 China started building four units of this design. The NRC provided construction permits for the AP1000 in the United States in 2011 and 2012. However, independent groups, including the Union of Concerned Scientists and Friends of the Earth, challenged the viability of the containment structure to withstand 9/11 types of attack. While the Chinese State Nuclear Power Technology Corporation moved ahead to build four AP1000 reactors as originally designed, activities in the United States were delayed as a new design for the containment structure was developed. The Chinese reactors were connected to the grid in the summer of 2018. China then initiated development of similar reactors of their own design based upon the AP1000 for their ongoing nuclear energy program. Construction delays, including the discovery that some welding certifications were falsely approved, led to Westinghouse filing for Chapter 11 bankruptcy in March of 2017. Plans to construct two of the AP1000 reactors in South Carolina were canceled in July of 2017, while plans to construct two reactors in Georgia continued. These are known as Vogtle 3 and Vogtle 4.

A major factor in the Georgia delays and cost overruns were miscalculations made by Westinghouse regarding the ambitious modular approach to construction that was used for the first time in this project. They sought to build prefabricated sections of the plant at remote locations and then ship these components to the construction site for assembly. Despite the untested nature of these plans, Westinghouse projected aggressive timetables for implementation. However, they misjudged both construction and regulatory factors.

Major delays resulted from their reliance on The Shaw Group to build sections of the reactor at their factory in Lake Charles, Louisiana, where the workers were unable to meet the stringent requirements of the NRC or maintain required quality-control standards. It was also reported that Shaw managers ordered employees to cover up incidents, for example, a sub-module having been dropped and damaged. Additionally, components were labeled improperly and paperwork went missing or was illegible.

A modular section from the Lake Charles facility was delayed for more than eight months because of missing signatures. The section was one of 72 modules fused together to hold nuclear fuel. The 2.2-million-pound unit was installed more than two years behind schedule.

The Lake Charles facility was sold in 2013; the new management took two years to establish acceptable production schedules. CB&I, the new owners of the plant, claimed that subsequent delays occurred due to Westinghouse making "several thousand" technical and design changes after work had started on various components.

While all references to the manufacturer are stated as Westinghouse Electric Corporation (WEC), the reality of ownership of that enterprise is quite complex. WEC was owned in 1999 by British Nuclear Fuels Limited and was then sold to Toshiba in 2006. After filing for bankruptcy in 2017, WEC was acquired by Brookfield Business Partners. From 1999 until 2023, WEC acquired seven other companies, including Mangiarotti (a heavy components manufacturing company), CB&I Stone and Webster, and Rolls-Royce nuclear services division. It is probably the case that these changes and acquisitions disrupted operations.

The 2017 start date for the reactors slipped to 2024 with cost mounting to $36.8 billion. However, the requirements for a different containment structure were issued after construction started and contracts signed. There were also significant problems with workforce performance, supply chain deliveries, and a change in management due to the bankruptcy. Nevertheless, the Vogtle reactors will be the first new operational reactors in the United States in three decades.

In August 2022, the NRC authorized Southern Nuclear to load fuel and begin operation at Vogtle 3. In January 2023, unexpected vibrations in the cooling system led to further delays. Operation began July 31, 2023. The two AP1000 reactors at the Vogtle site will be the first new reactors to be introduced into the US grid in 30 years and the only reactors with Generation-III safety features. Four reactors of this design are operating in China and setting a new standard for 1000-MW electric power generation. Having operational AP1000 reactors in the United States will lend credibility to Westinghouse's efforts to market this reactor in Central and Eastern Europe.

In the post-Chernobyl era, nuclear safety was widely discussed and analyzed. In the United States, Westinghouse was the first to pursue development of a nuclear reactor design with passive safety features and modular construction. As the difficulties that they encountered became known, developers turned to the possibility of manufacturing reactors that were much smaller than the AP1000. The Westinghouse 2017 bankruptcy was a stunning development that caused many to rethink strategic approaches to post-Chernobyl nuclear power development. It was seen that smaller reactors could be manufactured and transported more easily and assembled into larger power plants by connecting multiple small units. Further, successful factory production of small modular reactors could reduce costs and enhance quality control and safety.

A third advantage of small modular reactor implementation was the potential of having a footprint for the resulting power plant that could fit in the same area as the fossil fuel plant being replaced and attaching to the same grid connections that had been used by the fossil fuel unit.

The potential viability of small modular reactors was an idea that motivated scientists, engineers, and investors worldwide. This goal has been pursued since the beginning of the twenty-first century by research and development groups in many countries. Enthusiasm and optimism for the development of safe, economical, and transformative reactors that could easily replace carbon-emitting fossil fuel plants has stimulated investment by government agencies and private entities in many locations worldwide to actualize such reactors. Some of those that are Generation III are discussed here; Generation IV reactors are covered in Chapter 23.

While both Generation-III and Generation-IV reactors implement passive safety and modular fabrication, the Generation-III reactors use water as a coolant while Generation-IV do not. We recall that water must be kept under pressure to attain a temperature suitable to engage in effective transfer of heat from the core to the steam generator, thus requiring pressurizing technology and containment in a reactor's design. Water also has the potential of initiating a steam explosion (as at Chernobyl) or participating in oxidation reactions with the metal that encases the fuel thereby releasing hydrogen which could explode (as at Fukushima).

These factors persuaded many to opt for alternative coolants such as molten salts, liquid sodium, liquid lead, or helium gas; others used water because of the vast experience that had seen safe use of that technology.

Small Modular Reactors (SMRs)

United States

At the time of the writing, the company NuScale was the most likely to be the first to succeed in building a Generation-III water-cooled reactor (Figure 15.1). NuScale originated at Oregon State University from their joint research with Idaho National Laboratory funded by the US Department of Energy from 2000 to 2003. The research continued at Oregon State University, leading to the founding of NuScale in 2007.

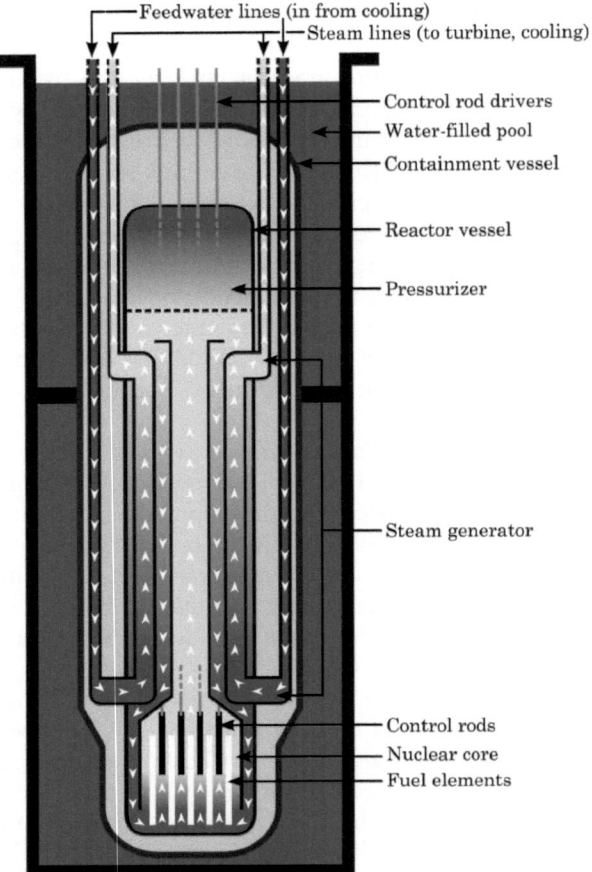

Figure 15.1 *NuScale reactor—a pressurized water reactor (PWR) that operates entirely through convection and gravity without the use of pumps.*

Reproduced courtesy of Alex Wellerstein (Stevens Institute of Technology, Hoboken, NJ, USA).

NuScale sought certification of their design beginning in 2008. When certification was received in 2023, it was the first of a new reactor design in more than 30 years. Achieving this breakthrough was possible because water-cooling technology was so well known.

The development of NuScale was interrupted in 2011 when its main investor, the Kenwood Group, was investigated by the US Securities and Exchange Commission (SEC) and found guilty of operating a Ponzi scheme. While NuScale finances were not associated with the Ponzi scheme, the funds of Kenwood were frozen by the SEC, which caused NuScale to downsize at an inopportune time.

NuScale went through most of 2011 in a precarious financial state resolved only when the Fluor Corporation, a major investment company that had been a shareholder in Westinghouse, saved the company. Fluor provided financial backing to NuScale in agreements that included rights for Fluor to construct NuScale-based power plants.

NuScale was able to secure $226 million in cost-sharing funds from the US Department of Energy in 2013 as well as $217 million in matching funds from the US Department of Energy in 2014.

NuScale entered an energy development program with utilities in the north-west United States for an initial plant to be built in Idaho. In August 2020, the NRC issued an approval of NuScale's safety design. Certification of the full design was approved by the NRC in January of 2023—the first approval of a new design in the United States in more than 30 years.

NuScale signed a contract in February of 2022 to build a plant in Poland and in December 2022 signed an additional contract for a plant in Romania. These would be the first small modular reactors (SMRs) in Europe and represented a breakthrough for the export of US nuclear technology.

SMRs also contribute to regional energy security and decarbonization goals. The war in Ukraine has placed a strong focus on the dominant role played by Russia in the energy sector of Eastern Europe, which the United States has been seeking to counter. Russia built 15 reactors in Ukraine prior to 2022. Both NuScale and Westinghouse have received support from US diplomatic representatives in Eastern Europe to counter that Russian market dominance.

The technology that NuScale is implementing is a creative break from the reactor designs dating back to the Manhattan Project. The 1000-MW size has become standard due to economies of scale relating to overall infrastructure that includes turbine use, cooling towers, and connections to the grid. To produce power in the 1000-MW electric range, NuScale employs a dozen small reactors that operate in unison in contrast with the universal practice of drawing power from a single large core. The design of the core module must meet transport requirements. Given that the basic modules must be transported by a truck or train in a container, the dimensions of the container determine the maximum size of the reactor. Since regulations in the United States specify that a truck can be no longer than 70 feet in length and 10 feet in width, it is unsurprising that the NuScale basic reactor is 65 feet in length and 9 feet in diameter.

The core is placed into this cylindrical metal containment structure at the bottom with pressurizers, tubes for water cooling, and heat exchangers packed into the space above. There are no pumps for external cooling since passive cooling is activated and maintained by gravity and nature's dynamic cooling currents. If there is overheating, then gravity will draw down cool water and maintain a flow of heated water.

The size constraint of the cylinder determines the maximum amount of energy that can be produced—which NuScale engineers determined to be 77 MW electrical. To meet the challenge of having a power station that provides approximately 1000 MW electrical, 12 basic units are tied together providing an output of 924 Megawatts. This NuScale Power Station design is sunk into the ground and filled with water. Each unit can operate independently and is connected to its own digital control panel in the operator room. Individual units can be disconnected for maintenance and refueling without interrupting the operation of the other units. The water in the containment pool provides additional emergency cooling, while the below ground-level placement of the reactor provides protection from earthquake damage and possible attack by terrorists using an airplane, as in 9/11.

The full power station requires connections for this power plant to generators and the use of cooling towers to bring the steam back to its liquid state. The output electricity can be connected to the grid in about 35 acres, which is an area less than or equal to the footprint of a coal or gas power plant.

Inflation during 2022 raised questions on the viability of NuScale's development plans. Important construction components such as copper wire and steel piping experienced cost increases of 32% and 106%, respectively. Increases in interest rate also led to increases in construction costs. The estimated costs of building the power plant in Idaho rose by more than 50%, which translates into an increase in the cost of energy from $58 to $89 per MW/hr. The clients for this power are a group of municipalities in the region known as Utah Associated Municipal Power Systems (UAMPS). These increases in projected costs forced the November 2023 cancellation of the NuScale contract in Utah. As of the writing of this book, plans for implementation of this technology in Romania and in South Korea are still moving forward.

Given its leadership in manufacturing PWRs it is surprising that Westinghouse did not initiate a Generation-III+ SMR development program until May of 2023. At that time, it announced plans to build the AP300, a scaled-down version of the AP1000. Westinghouse was optimistic that AP300 development and licensing could be expedited due to experience with the AP1000.

In parallel with Westinghouse as the lead producer of PWRs, we have seen GE–Hitachi as the dominant manufacturer of BWRs, who have built more than 20 across the world. Given that there is a significant advantage in developing an SMR using a known technology that has received licenses for construction from the NRC and is known and trusted in the United States and in many other countries, it is unsurprising that GE–Hitachi decided to pursue development of an SMR that uses boiling water in the design.

The GE–Hitachi entree in the SMR arena is the BWRX-300. This is a smaller and simplified version of their licensed Economic Simplified Boiling-water Reactor (ESBWR) and is the only BWR design that is being pursued for implementation as an SMR. The transformation reduces the plant size by 90% and the power output by two thirds. This 300-MW electric reactor will produce steam inside the pressure vessel—a design strategy that greatly reduces the complexity and size of the plant.

GE–Hitachi has been successful in their marketing of the BWRX-300. They are engaged with the Tennessee Valley Authority (TVA) for construction of this reactor near Oak Ridge, Tennessee. While outside of the United States they have several contracts signed or moving toward finalization, their first contractual engagement was with Canada's Ontario Power Generation for a unit at the Darlington Nuclear Generating Station, which is 70 km east of Toronto. Other contracts are being finalized in Poland and in Estonia.

United Kingdom

The United Kingdom has not been an active player in the development of nuclear reactors until a recent initiative by Rolls-Royce. Despite advice from some in the United Kingdom that Rolls-Royce should seek collaboration with GE–Hitachi and obtain a license to manufacture the BWRX-300, they have decided to build a modular PWR of their own design. Rolls-Royce has not had previous experience in commercial nuclear power construction for civilian use, although they did succeed in building the nuclear propulsion systems for the British nuclear submarines.

In March 2022 Rolls-Royce submitted their reactor design for review by the regulatory agencies in the United Kingdom. The company plans to build its first unit in the early 2030s and to construct ten in total by 2035. While the generally accepted specification for the power output of a small nuclear reactor is ≤ 300 MW electrical, the Rolls-Royce design is for about 450 MW electrical. However, they assert that their reactor will be factory built with modules that can be transported by truck, train, or ship. A PWR is planned. While they are in an early stage of development, they are considering potential sightings for these reactors with a focus on locations in Wales, although Cumbria is the initial location.

France

France has been the most aggressive advocate of nuclear energy in the world, with more than 70% of its electrical energy generated by nuclear power plants. In recent years the French government implemented a plan to reduce their use of nuclear energy to 50%, but that decision is being reconsidered.

In 2022, President Emmanuel Macron succeeded in having the European Union include nuclear power in programs aimed at reducing global warming. This was done despite opposition from Germany. Also, in 2021 as part of the France 2030 plan, Macron included the promotion of small nuclear reactor production with an allocation of one billion Euros. His vision is to have France be a leader in the export of nuclear technology. As of early 2023, there has been no publication of the design features for the proposed French SMRs.

Further Reading

"Advanced Nuclear Power Reactors", *World Nuclear Association* (last modified April 1, 2021). https://world-nuclear.org/information-library/nuclear-fuel-cycle/nuclear-power-reactors/advanced-nuclear-power-reactors.aspx

"Generation-III Nuclear Reactors," *Wikipedia* (last modified December 24, 2024) https://en.wikipedia.org/wiki/Generation_III_reactor

These are both reviews of Generation-III nuclear reactors.

16

Fuel, Waste, and Radioactivity

All the waste in a year from a nuclear power plant can be stored under a desk.

— Ronald Reagan[1]

Nuclear Waste and Radioactivity

Nuclear power is widely misunderstood and there are no aspects of it more misunderstood than two words that communicate danger and fear: waste and radioactivity (Figure 16.1).

The common refrain "What about the waste?" has persisted despite the development of robust ways to handle the spent fuel, or waste, of commercial reactors—ways that have resulted, globally, in a perfect safety record. As often as the question is asked in earnest, it is also asked by those who oppose the use of nuclear energy to cast doubt on its safety.

Fear of nuclear waste plays an inordinate role in shaping popular opposition to the use of nuclear power. This excessive fear of nuclear waste is due, in part, to guilt by association with the destructive capacity of nuclear weapons. That fear is exacerbated by the intangible nature of nuclear radiation which is invisible to human eyes.

As laid out in this chapter, scientific evidence does not support rejection of nuclear energy based on the risk of radiation from reactor waste.

The generation of electricity from a typical 1,000-megawatt (MW) electric nuclear power station, which can supply the needs of more than a million people, produces only three cubic meters of vitrified high-level waste per year. In comparison, a 1,000-MW coal-fired power station produces approximately 300,000 tons of ash and more than six million tons of carbon dioxide every year.

The deleterious impact on health of fossil fuels includes deaths that can be attributed to air pollution. The death toll from the burning of fossil fuels—in power generation, transportation and industry—total more than 3.5 million premature deaths annually. This total rivals all the deaths annually from murders, wars, and terrorist attacks combined.

The comparison with nuclear waste related deaths is stark. The annual total of deaths attributable to nuclear waste is, and has been, zero! The inability of this comparison to

---

[1] Ronald Reagan, February 15, 1980. *The Burlington Free Press*, p. 14.

Nuclear Energy. Edward A. Friedman, Oxford University Press. © Edward A. Friedman (2025). DOI: 10.1093/9780198925811.003.0016

Figure 16.1 *Radiation warning sign.*
Reproduced from Wikimedia Commons. (2006). https://commons.wikimedia.org/wiki/File:Radioactive.svg.

gain traction in energy policy discourse is puzzling. Part of the reason for this lack of perspective lies in the fact that nuclear waste exists in a fog of misunderstanding. It is important to provide clarity on the nature of nuclear waste and the basic facts about the emanations that we refer to as "radiation."

A response to the question, "What about the waste?" must start with an answer to the question: "What **is** the waste?" followed by spelling out the connection between waste and the impact of the resulting radiation.

Many even question the use of the term "waste" in describing the material extracted from a reactor following the use of fuel to generate energy! What is seen as waste in one context can be seen as fuel in another context!

Nuclear waste is a broad category covering any radioactive material produced during any nuclear-related activity, including commercial power production, nuclear medicine, nuclear fuel reprocessing, uranium mining, experimentation at a research reactor, etc. Categorized by the level of hazard, 90% of nuclear waste by volume is considered "low level," such as worker radiation suits, gloves, or used tools that can be easily managed. Around 7% is considered intermediate level, such as irradiated reactor components that have been removed and replaced during maintenance. Just 3% of the waste is considered high-level waste, though 95% of the total radioactivity emitted by nuclear waste is contained within this relatively small portion. It is this 3% of waste, the high-level waste, that has defined nuclear waste in the popular imagination.

Before it became high-level waste, it was nuclear fuel. Used nuclear fuel is considered as waste after it has served its purpose and been removed from the reactor—not in some altered, sludge state that could stain your clothes, and not in oozing barrels loosely sealed, but in a nearly identical form to that in which it entered: as solid, ceramic, uranium fuel pellets, neatly contained within sealed metal tubes grouped into larger assemblies (Figure 16.2).

There is only an infinitesimal difference between the fuel that has undergone fission and residue that remains. The fission process has replaced uranium atoms with lighter elements that carry off energy, which produces heat. Some of these lighter elements are

Figure 16 2 *Dry storage of spent nuclear fuel in concrete casks.*
Reproduced from Nuclear Regulatory Commission. (2008). Wikimedia Commons. https://commons.
wikimedia.org/wiki/File:Nuclear_dry_storage.jpg.

radioactive, as are some atoms that have absorbed neutrons to create radioactive elements that are heavier than uranium atoms. Because a significant amount of fissile material remains within the spent fuel, the term "waste" could be considered a misnomer. Light-water reactors cannot efficiently use their spent fuel anymore, but technologies such as fast breeder reactors could generate tremendous amounts of additional energy from this material.

The fuel used in nuclear reactors is unlike any other fuel used for energy production. Rather than use the process of combustion to create heat (as in fossil fuels and some uses of hydrogen) or the movement of matter to spin generators (as in hydro or wind power), nuclear fuel is designed to enable the release of the energy contained within the nucleus of an atom. The process of fission, by which nuclear fuel releases energy, converts a miniscule amount of mass into a vast amount of energy in accordance with Einstein's equation, $E = mc^2$ (where E stands for energy, m for mass, and c for the speed of light). Doing so produces a small volume of waste.

Generating continuous heat in a reactor requires a nuclear chain reaction to take place. Prompting this chain reaction is the release of neutrons with each occurrence of fission, or the splitting of a uranium atom, which in turn produces more neutrons that trigger additional fissions. To overcome the loss of neutrons that sustain the fission reactions from their absorption in water coolant (which thereby interrupts the chain reaction), the uranium fuel is "enriched" to increase the proportion of the fissile uranium-235 (U-235) isotope. Enrichment means that the percentage of U-235 in the fuel is increased. While the proportion of U-235 is 0.7% in uranium from a mine, enriched uranium used in pressurized-water reactors (PWRs) constitutes U-235 levels of 3–5% of the fuel. Uranium enrichment is strictly regulated by the NRC and other national regulators, the UN through the International Atomic Energy Agency (IAEA), and other multilateral agreements.

Uranium is abundant in the Earth's crust. The element was first identified in 1789 by Martin Heinrich Klaproth in Germany. He discovered it when experimenting with pitchblende ore. Uranium mines are found in many countries. Kazakhstan, Canada, and Australia are the top producers of uranium, accounting for 68% of world production.

Historically, uranium was mined with crude methods that involved open pits with minimal protection for workers, prior to the strict regulations and oversight which now govern the mining of uranium. Uranium production had a significant, albeit localized, environmental impact.

Today's methods of mining uranium have advanced to include in situ and other methods that do not disturb the surface soil and have significantly reduced the environmental impact of these mines. In-situ mining involves pumping a chemical through a borehole into the ground that dissolves the ore, which is then pumped out through a second borehole.

After being mined, uranium ore must be prepared for enrichment through a series of chemical conversions to create uranium hexafluoride, the form in which it is most easily enriched. After enrichment, it is fabricated into hard, ceramic uranium oxide fuel pellets. The configuration in which it is loaded into the reactor differs between plant designs. In light-water reactors, however, these fuel pellets are stacked into thin metal tubes, called fuel rods, and arranged into bundles, called fuel assemblies. Many such fuel assemblies are carefully placed within the reactor core, swapping in for a portion of the used assemblies at carefully coordinated refueling stages every 12–18 months. Other designs exist, such as the Canadian CANDU reactor, which opt for smaller fuel bundles rather than large assemblies and have the feature of on-power refueling that doesn't require set outages. Owing to the high energy density and low cost of uranium fuel, fuel costs are not a significant cost-driver of nuclear power plants.

Many types of fuel were tried during the experimental phase of nuclear energy during the 1950s and early 1960s. Dissolved fuels within the reactor coolant, coated fuels, fuels with different enrichment levels, and more made their way into experimental reactors. The global use of ceramic uranium oxide fuel pellets, originating from the United States' first PWR at Shippingport, followed the widespread adoption of PWRs and boiling-water reactor (BWR) technology. As the types of nuclear reactors diversify once more, so too will the types of fuel used. There are several potential configurations and enrichment levels of nuclear fuels, resulting in characteristics suitable for different types of reactors. Already, some reactors, such as heavy-water reactors, use unenriched, or natural, uranium fuel, owing to a lower neutron absorption by their moderator and coolant. Some fuels in use today are also manufactured through the reprocessing of used fuel.

For example, such reprocessed fuel powers much of the French nuclear fleet, giving the country a closed nuclear fuel cycle.

The use of improved fuel designs is replacing the fuel pellets used in most reactors currently operating. This advanced fuel is TRISO, which promises to contain its radioactive fission products within a special coating, qualifying it as "accident tolerant." TRISO fuel can be used in several different types of nuclear reactor designs.

Safe TRISO Fuel

The name TRISO is derived from the technical description of the structure—a tri-structural isotropic particle. TRISO has the uranium fuel encapsulated in a small ceramic chip about the size of a poppy seed. These particles are about 0.04 inches (1 mm) in length and consist of the uranium and carbon surrounded by layers of ceramic. These small particles are then packed into billiard ball-sized spheres called "pebbles." A pebble bed nuclear reactor in China uses spheres containing 12,000 TRISO particles.

TRISO fuel is resistant to neutron irradiation, corrosion, and the highest temperatures attainable by the nuclear reactors. Each particle acts as its own containment system allowing it to hold the radioactive decay products produced during the production of energy by the chain reaction of the U-235. The TRISO particles cannot melt, thereby providing safe containment of the radioactive waste produced by the fission process.

TRISO fuel was first developed in the United States and in the United Kingdom in the 1960s. Its current formulation emerged from an initiative in the US Department of Energy in 2002 in anticipation of the need for fuel that could withstand high temperatures in advanced design reactors. In 2009, TRISO fuel set an international record by achieving 19% maximum burnup during a three-year test at Idaho National Laboratory. This is three times the burnup that can be achieved in standard light-water reactors.

The irradiated fuel was then exposed to more than 300 hours of testing at temperatures up to 1800 °C (more than 3,000 °F). These tests exceeded the predicted worst-case accident conditions for high-temperature gas reactors and showed no minimal damage of the particles with full retention of the fission products.

In the United States several companies have been licensed by the government to produce TRISO fuel. These include X-Energy, the Ultra Safe Nuclear Corporation, and BWX Technologies. China independently produces TRISO fuel for its pebble bed High-Temperature Gas Reactor.

Because TRISO enables a high temperature output for advanced reactor designs, it has the capacity to sustain the use of nuclear energy for applications of direct process heat, thereby replacing fossil fuels in applications such as metals processing, desalination, and hydrogen production. TRISO fuel use is also planned for Generation-IV gas-cooled reactors and molten salt reactors.

The Enrichment Bottleneck

Uranium extracted in mining consists almost entirely of uranium-238 (U-238),with less than 1% fissionable U-235. A low-enriched fuel, at about 5% U-235, is far from bomb-grade level of 90%, yet still requires significant enrichment. For fuels like TRISO, the enrichment needs are even greater (up to 20% enrichment). Today, global uranium producers (miners and fuel fabricators) far outnumber global uranium enrichers. Enrichment is dominated by a small group of companies operating eight countries: Tenex (Russia, owned by Rosatom), Urenco (US, Germany, Netherlands, UK), Orano (France, formerly Areva), and CNNC (China).

Enrichment capacity in the United States and Europe is insufficient to meet the need for enriched fuel, resulting in about a third of the enriched fuel for the US nuclear fleet coming from Russian enrichment facilities. This portion is even higher in Europe. There is much debate over how quickly the United States and Europe can ramp up enrichment—a topic that soared in relevance after the Russian invasion of Ukraine in early 2022. There was a surge of concern over energy security as Russia showed its willingness to weaponize its energy exports to put pressure on other countries. Among enrichment companies operating in the United States and Europe, there is general agreement that governments must support and backstop the large investments needed to increase uranium enrichment capacity to supplant Russian supply. In late 2023, support for the Sapporo 5 enrichment group, which includes the United States, Canada, Japan, France, and the UK, was announced, with $4.2 billion earmarked for increasing enrichment capacity.

The line demarcating low-enriched uranium and high-enriched uranium is 20% enrichment. Several new proposed reactor designs make use of higher-reactivity fuel rather than the standard 5% enriched fuel used in today's LWR fleet. This uranium, enriched to 5–20%, is known as High-Assay Low-enriched Uranium, or HALEU.

Fueling reactors will be a significant challenge as more countries turn to nuclear power. Designs that use innovative fuels will face particular hurdles, with HALEU production as a choke point. Those who are engaged in fabricating fuels for evolving reactor designs are challenged by the task of incorporating enough enriched uranium.

Radioactivity Basics

Nuclear energy technology is unique for many reasons. While nuclear energy can provide scalable, round-the-clock power without carbon emissions, it is viewed as dangerous and subjected to severe regulations, international treaties, and scrutiny by political action groups because of its association with radiation and its relationship with nuclear weapons.

The possibility of nuclear accidents strikes fear into people because of the potential release of radiation—an invisible cancer-producing agent even at low doses. The reason that nuclear weapons are considered uniquely dangerous, apart from their unprecedented explosive force, is their ability to contaminate swaths of land with radioactive fallout. Thoughts about nuclear waste are inextricably tied to thoughts about nuclear weapons.

Next to a mushroom cloud, the image most strongly associated with radiation is the trefoil around an inner circle. This universal symbol for radiation is often associated with danger and fear of invisible forces. The relationship between nuclear radiation and nuclear weapons is hard to ignore. The fact that the nuclear weapons attacks on Hiroshima and Nagasaki caused more than 100,000 deaths deeply influences public reaction to any source of radiation in society to the present day. The invisibility of nuclear radiation further enhances its ominous nature. Yet, a quantitative assessment of harm done by radioactive waste from nuclear reactors pales next to the millions of deaths caused annually by air pollution from coal-fired power plants.

Some will find it astonishing that there are no known cases of people dying from mishandling of spent nuclear power reactor fuel. This is true in the many countries that use nuclear reactors for production of energy.

Radiation Dose Explained

Radiation is a catch-all term that refers to the impact on human beings of electromagnetic energy in the form of X-rays and gamma rays as well atomic particles including electrons, neutrons, and the nuclei of helium atoms with the electrons referred to a beta radiation and the nuclei of helium atoms referred to as alpha rays (Figure 16.3).

The nomenclature is confusing since the use of the term "rays" is misleading when referring to the electrons, neutrons, and helium nuclei, which are particles. X-rays and gamma rays are both forms of electromagnetic energy with gamma rays having shorter wavelengths than X-rays.

The impact of all these forms of radiation is summed up with a common unit called a sievert, which is a measure of how much damage is done to human tissue when a given amount of radiant energy is absorbed. However, given amounts of neutron and

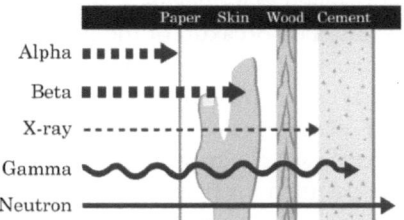

Figure 16.3 *Different forms of radiation can penetrate or be blocked by different substances.*

Reproduced courtesy of Alex Wellerstein (Stevens Institute of Technology, Hoboken, NJ, USA). Adapted from Nuclear Regulatory Commission. (2010). Wikimedia Commons. https://commons.wikimedia.org/wiki/File:Penetrating_Power_of_Radiation.svg.

alpha radiation energy are seen to incur greater damage than an equivalent amount of electromagnetic energy and electron energy and are assigned higher sievert values.

A baseline for discussion of radiation is the amount of radiation that we are all exposed to all the time. This is the "background" radiation that is due to low levels of radiation present in the world around us. Everything from bananas and marble to radon gas that enters our basements from the ground adds to this background. The universe in which we live has particles traveling through space at high velocities that strike the Earth's atmosphere. The resulting radiation is referred to as cosmic radiation, which also contributes to background radiation. While background radiation varies by location, depending on altitude and materials in the ground, there is an average number used for those who live in the affected areas of Europe, including the former USSR. Note that the average background figures for the United States are higher than the worldwide averages. While the technical value for this worldwide average is 2 millisieverts (mSv) per year, we can simply refer to this as B for background radiation and consider other radiation relative to that amount.

The next relevant number is the radiation that the average person receives from medical X-rays, CT scans, etc. This turns out to be about equal to the background radiation. Therefore, inescapable human exposure is 2B (or two times the background radiation dose) for one year.

For workers who are normally exposed to radiation, the limit for an acceptable dose in a five-year period is 50B. The United Nations Scientific Committee on the Effects of Atomic Radiation (UNSCEAR) has estimated that the additional chance of dying of cancer due to radiation exposure above 50B is about 4% per sievert. On the high end, a dose of 2,500B would kill half of those exposed within a month.

Our focus is on nuclear waste from power reactors. Nuclear waste also occurs from uranium mining, weapons testing, and processing of radioactive materials for medical applications. Generally, different programs are employed to deal with each of these nuclear categories.

The generation of nuclear energy produces radioactive material via two mechanisms. One is via the fission process in which neutrons cause the uranium nucleus to split into two fragments that are sometimes unstable and the other is when neutrons are absorbed and create heavier unstable nuclei.

The fission of U-235 occurs when that nucleus absorbs a neutron forming the unstable intermediate form of U-236. The breakup of the intermediate U-236 is a statistical process with numerous outcomes. This breakup yields two lighter elements along with the release of free neutrons that carry substantial amounts of kinetic energy.

What follows is a nuclear physics calculation that may appear intimidating to a non-scientist, yet it is a matter of simple arithmetic. A few examples will illustrate the straightforward nature of the calculation. If we start out with U-235 plus the neutron that precipitates the fission, we have 92 protons and 143 neutrons in the U-235, plus the single projectile neutron prior to the fission for a total of 144 neutrons. The fission itself can result in many possible outcomes with the two resulting nuclei most likely having proton numbers in the 50s and 30s. Barium with 56 protons and krypton with 36 protons being an example. We see in this example that krypton has 53 neutrons and barium has 88 neutrons. The neutrons present in these two nuclei plus the three free neutrons add up to the number in the U-235 nucleus that split, which was 144.

→ U-236: Barium 144 + krypton 89 + 3 neutrons

→ Proton count: Uranium 92 = krypton 36 + barium 56

→ Neutron count: Uranium 144 = krypton (89 − 36 = 53) + barium (144 − 56 = 88) + 3 free

Another example of the fission of U-235 is the process that yields xenon and strontium. Here the number of protons add up to 92 with xenon contributing 54 and strontium 38. The neutron balance sees 86 neutrons in xenon and 56 neutrons in strontium with two free neutrons again totaling 144.

→ U-236: Xenon 140 + strontium 94 + 2 neutrons

→ Proton count: Uranium 92 = xenon 54 + strontium 38

→ Neutron count: Uranium 144 = xenon (140 − 54 = 86) + strontium (94 − 38 = 56) + 2 free

One of the most prominent fissions of U-235 produces cesium-137, which has 55 protons, and rubidium, which has 37 protons. It is left to the reader that the neutrons add up to 144 when four free neutrons are considered.

→ U-236: Cesium-137 + rubidium 95 + four free neutrons

These fission fragments are themselves often unstable and emitters of radioactivity. Prominent in this category is cesium-137, which emits gamma radiation and decays with beta emission (a transition in which a neutron in its nucleus transforms into an electron and a proton). This process has a half-life of 30 years. The term "half-life" refers to the amount of time that passes before a given level of radiation emission is reduced to a level that is one-half as intense.

It is transformed fuel that constitutes nuclear waste. After removal from a nuclear reactor most of the waste that is prominent after 30 years consists of cesium-137 and strontium-90, which are each produced in about 6% of fissions and have half-lives of about 30 years. These elements are most prominent in spent fuel in the period from several years to several hundred years after use.

Strontium has 38 protons and 52 neutrons and is paired with xenon with 54 protons and 90 neutrons plus two free neutrons adding again to 92 protons with 144 neutrons. Strontium-90 was present in the fallout from nuclear testing conducted in the atmosphere over the Nevada desert. Because it has chemical properties similar to calcium, it was absorbed in the teeth of young children in measurable amounts and became the centerpiece of the atmospheric anti-nuclear testing campaign in the United States.

The half-lives of radioactive elements emanating from most U-235 fissions are short, ranging from seconds to tens of years. Examples are barium-141 at 55.72 seconds;

krypton-82 at 5.84 seconds and lanthanum-146 at 6.27 seconds. The three radioactive isotopes that have significant biological consequences ^{131}I, strontium-90, and cesium-137 have half-lives of 8.05 days, 29 years, and 30 years, respectively. We see that after 200 years the impact of these radioactive elements is inconsequential.

After cesium-137 and strontium-90 have decayed to low levels, a significant proportion of the radioactivity in spent fuel comes not from fission products but from heavy radioactive nuclei that were formed when the uranium absorbed neutrons and formed isotopes of plutonium and other heavy nuclei. This second mechanism for the generation of radioactive elements in nuclear waste referred to as the production of transuranics or nuclei with more nucleons than uranium.

At the time when the fission of uranium was discovered in 1939, the number of known elements was 90, with uranium being the atom with the largest number of protons and electrons. The intensified study of uranium bombardment with neutrons led to the discovery of plutonium. Plutonium-239 has 94 protons and 145 neutrons. It is formed when U-238 absorbs a neutron and then undergoes beta decay twice with neutrons changing into protons. Plutonium 240 is also produced. Plutonium-239 has a half-life of 24,100 years and plutonium-240 a half-life of 6,560 years. The half-lives of these radioactive transuranic elements are in sharp contrast with those of the radioactive fission fragments found in nuclear waste.

A small percentage of the radioactive waste originating in the fission process also has extremely long half-lives. These consist of seven isotopes:

→ Technetium-99 with a half-life of 211,000 years
→ Tin-126 with a half-life of 239,000 years
→ Selenium-79 with a half-life of 327,000 years
→ Cesium-135 with a half-life of 1.33 million years
→ Zirconium-93 with a half-life of 1.53 million years
→ Palladium-107 with a half-life of 6.6 million years
→ Iodine-129 with a half-life of 15.7 million years

The amount of radiation emanating from these isotopes is extremely small.

Nuclear waste can be divided into three groups according to somewhat arbitrary categories of low, intermediate, and high levels of intensity. The vast majority of the waste (90% of the total volume) is composed of lightly contaminated items such as tools and clothing. This material contains only 1% of the total radioactivity. In contrast, high-level waste constitutes 3% of the total volume and 95% of the total radioactivity. The remaining waste, categorized as intermediate level, constitutes 7% of the total volume and 4% of the total radioactivity.

Ongoing monitoring of the status and trends regarding spent fuel and radioactive material including inventory and maintenance programs is conducted by the IAEA in accordance with the Joint Convention on the Safety of Spent Fuel Management and on the Safety of Radioactive Waste Management. These conventions were established by the IAEA in 1997 and gained the adherence or ratification of most member states in subsequent years.

At North America's largest nuclear facility, spent fuel (waste) from Ontario's nuclear fleet is stored at a single facility, named Ontario Power Generation's Nuclear Sustainability Services. If invited to tour the facility, one will find not a dingy prison struggling to detain the "mankind's deadliest garbage," but a silent, warm, immaculately clean warehouse—a warehouse engineered to survive direct plane crashes and which sports a 20-foot thick basement of reinforced concrete. In it stand evenly spaced rows of large, numbered, pearl-white containers. The containers consist of layers of stainless steel and high-density concrete, welded shut with a weld whose filler material alone weighs 70 pounds. Inside the fortified containers is nuclear energy's bogeyman, the waste: solid, dry, contained in corrosion-resistant zirconium alloy casings, and held in specially constructed racks. The shielding the containers provide is so robust that one can stand immediately next to one of the concrete-and-steel sentinels without receiving more than a background dose of radiation.

Walking through this warehouse, which despite housing 40 years of waste from the largest nuclear plant on the continent only takes a few minutes to cross, one is struck by the security of it all, by the inconceivable series of events it would take to disrupt the containers sitting quietly there, to dislodge the solid fuel that is incapable of "leaking" at all, and to proceed to expose the far-off public to any measurable threat.

There is no threat to health walking through the facility, as radiation is always closely monitored. Nevertheless, it is difficult not to find this facility ominous since there is, of course, risk when managing spent nuclear fuel. When first removed from the reactor core, the spent fuel is "hot" enough that even a brief exposure could be lethal for an unshielded worker. However, a thorough understanding of radiation and how to shield it, and the development of handling procedures and technology, have been able to avoid injury entirely. For instance, radiation is entirely blocked by a few feet of water, so fuel is transferred out of the reactor while remaining submerged, and it remains submerged in a spent fuel pool for several years until the radiation naturally reduces to a level at which it can be safely transferred to a dry storage container.

These pools are made of reinforced concrete that is several feet thick and which have steel liners. The water in the pool is typically 40 feet deep and serves to both shield and cool the rods. After about three to ten years in a cooling pool, the spent fuel rods are embedded in stainless steel canisters that are surrounded by concrete. These casks are typically certified for up to 40 years with options for renewal. In the history of the commercial nuclear sector, there are no known deaths or injuries from the handling of spent fuel.

Yucca Mountain Nuclear Waste Repository

The Yucca Mountain Nuclear Waste Repository was designated as the sole repository for U.S. spent nuclear fuel and high-level civilian waste in 1987. The site is on federal land adjacent to the Nevada Nuclear Weapons Test Site located about 100 miles northwest of Las Vegas. This proposal has met with fierce political opposition from public groups and political leaders in the state of Nevada. It has also received in-depth scrutiny for safety.

The site was designated in 1987 as the sole repository for nuclear waste by legislation. Action on construction was delayed because of controversy that involved court battles. In 2010, President Barack Obama signed an act that repealed the designation of Yucca Mountain as the sole repository for nuclear waste and required the Department of Energy to study alternative sites.

In 2019, the Trump administration restarted the Yucca Mountain project and sought approval from the Nuclear Regulatory Commission, which was rejected in 2021 citing numerous technical and regulatory deficiencies.

More recently, plans remain in limbo with political debate continuing in Congress without any resolution. Much of the controversy relates to a call for the repository to be effective for a period of 10,000 years. While it is difficult to guarantee safety for that long of a period from potential earthquakes and formation of volcanoes, it is also the case that, after a few hundred years, the spent fuel could be relocated or reprocessed. This inaction on creation of a national repository in the United States has required continuation of the use of storage at individual reactors in pools and casks.

France

France has had the most ambitious nuclear energy program of any country in the world, with 65–70% of its electrical energy provided by nuclear reactors. Its 56 nuclear reactors generate a large amount of nuclear waste, which has been subjected to a high degree of reprocessing. The recycling of spent fuel and vitrification and burial of high-level waste had been centralized at La Hague in the northwest tip of the country. France also contracted with Russia to reprocess a portion of their nuclear waste in an arrangement that has been discontinued. The capacity of this facility, which began operations in 1976, is approaching its limits. As of the writing of this book, France is pursuing plans for expansion of their ability to process and store nuclear waste. High on their agenda is a proposal to build a deep storage facility in eastern France that would hold waste in clay formations at a depth of 1640 feet.

United Kingdom

As of mid-2024, the United Kingdom has nine operating nuclear reactors and two under construction, and given that 36 reactors have been shut down, it has accumulated large amounts of nuclear waste. Britain has centralized management of nuclear waste at Sellafield, a two square mile site on the Cumbrian coast. This location also manages waste from Britain's nuclear weapons program. In addition, the United Kingdom has accepted waste for processing from EU countries including Italy, Germany, and Sweden through agreements with Euratom. While reprocessing waste, Britain has produced a stockpile of over 300,000 pounds of plutonium. The future of that stockpile has yet to be determined.

While high-level waste at Sellafield has been vitrified and placed in secure silos, some lower-level liquid waste is reported to be in storage containers that are leaking. The authorities at Sellafield have not been entirely forthcoming about the history and nature of these leaks. Also troublesome are reports published by investigative reporters from *The Guardian* that computer systems at Sellafield have been penetrated by hackers from Russia and China. These revelations have led to the severing of nuclear related collaboration between the United Kingdom and China.

Britain is engaged in early stages of developing deep geological sites for nuclear waste disposal. Negotiations are underway to establish local community agreement for such facilities.

Germany

Germany built its first nuclear reactors in the 1970s and reached its peak implementation in the mid-1980s. It is the only country to walk away from nuclear technology—a process that began in 2009 and reached its conclusion in 2024. In that year, its last nuclear power reactor was closed leaving a landscape that is populated with 36 shuttered nuclear reactors.

Not only must Germany deal with long-term storage for the waste that was produced through years of nuclear reactor operations, but also they must deal with the waste that will be generated by the decommissioning of all the closed reactors. A proposal to use a salt mine in Gorleben met with protests from local citizens that led to the withdrawal of this suggestion. Legislation that was passed by the German Parliament in 2017 calls for extensive public involvement in the decision process regarding waste disposal plans. The timetable for the program that is eventually adopted will probably extend into the next century. Agreement on a plan is not envisioned before the early 2030s with the ensuing construction requiring a 20-year period and the actual transport of radioactive materials to this repository taking decades.

Russia

During the early days of the nuclear era, the Soviet Union exhibited a reckless disregard for the disposition of nuclear waste. Waste was disposed of in the oceans of the Arctic as well as the rivers and lakes of the Ural Mountains. This waste originated from the military as well as from public reactor programs. In the period 1946–1982, the Soviet Union was not alone in dumping radioactive materials into the oceans—a practice that was pursued by a dozen countries, including the United States and the United Kingdom. However, the Soviet practice far outstripped any other country with totals twice the cumulative amount from all other countries. It included 18 reactors from nuclear submarines, with two of the submarines still loaded with fuel. A nuclear icebreaker was also sunk along with barrels of low-level waste. While some dumping occurred in the Sea of Japan, most of the disposal took place in the Barents Sea and Kara Sea. Since the former is adjacent to the Norwegian Sea, there is concern that contamination can injure Norwegian fisheries.

A Convention on the Prevention of Marine Pollution by Dumping of Waste and Other Matter was adopted in 1972 as an international accord that today includes 87 countries. While Russia was among the original signers of this accord, it did not adhere to its provisions for many years. It was the breakup of the Soviet Union in 1990 that saw Russia begin to adhere to international norms.

A joint Russian–Norwegian cleanup effort in this region was aborted because of diplomatic tensions that arose from the Russian war in the Ukraine. In the Urals, Lake Karachay was identified as a depository of nuclear waste and was used for that purpose until 1957 at the time of the Kyshtym Disaster (discussed in Chapter 12). The lake was filled in with 10,000 hollow concrete blocks between 1978 and 1986 and with additional rock and soil in 2015 making it a near-surface permanent dry nuclear waste storage facility. High-level radioactive waste also contaminated the Techa River in the Urals around 1950. These Soviet-era waste events were kept secret both inside and outside the Soviet Union.

Currently, in the public sector, Russia operates 11 RBMK Chernobyl-type reactors, 11 VVER 1000-MW reactors, five VVER 440-MW reactors, four small PWRs, two liquid-sodium-cooled reactors, and one floating nuclear reactor. As of the time of writing, 11 reactors have been closed. Reprocessing and storage of spent nuclear fuel is centralized in Ozersk, the birthplace of the Soviet nuclear weapons program. This city in the Urals was originally a secret location that did not appear on any maps and its residents were not included in the Soviet census. Extreme secrecy still prevails in this nuclear processing headquarters for nuclear fuel and waste operations of Rosatom, the Russian organization that oversees this facility. Dangerous levels of radioactivity undoubtedly prevail in this city whose residents have chosen to live with high levels of exposure.

China

China began production of nuclear reactors much later than other countries with production of their first reactors in the 1980s and 1990s. The accident at Fukushima in 2011 precipitated a review in China of its nuclear energy program. After that review, China proceeded with particularly well thought-out programs for waste management in the context of systematic plans for Generation-IV reactor development.

Early in their development of nuclear energy, in the 1980s, China had successfully achieved operational success of their own centrifuge technology and entered indigenous production of enriched uranium. They also established reprocessing facilities that allowed production of plutonium for fuel and weapons. A centralized used fuel storage center has been built at Lanzhou Nuclear Fuel Complex in Northwest China. In addition, regional storage centers have been developed close to reactor sites. At these sites, vitrification of high-level waste is being pursued following an initial cooling period. The long-term plan is to build a geological storage complex in a remote area of the country with a timetable extending several decades into the future.

India

India has 23 operational reactors and an ambitious program for expanded reactor use with seven plants under construction in the mid-2020s. As of that time, one small reactor has been shut down. Waste storage and reprocessing plants are located at reactor sites with three national locations that can implement vitrification. Planning for development of a deep geological placement site is being pursued. There is an active program of extracting isotopes from reactor waste for use in industrial and medical applications.

Finland

In collaboration with the Swedish Nuclear Management Company SKB, Finland is preparing a site which will be the world's first repository for deep geological storage of nuclear waste that can be safe for at least 100,000 years. This facility will be ready for use in the mid-2020s timeframe. The location is Olkiluoto Island off the coast of Finland. Studies have been completed that found that the bedrock on this location was suitable for long-term storage. This repository will likely become a model to be emulated by other countries in the coming decades.

Decommissioning

Upon the termination of use, a nuclear plant disassembly and cleanup is implemented to allow reuse of the land that was occupied by the facility. These actions can be taken immediately or deferred usually for a period of 40–60 years. In the case of deferral, a process known as Safstor (safe storage) is implemented that ensures the safety of the radioactive materials. In the United States by the mid-2020s about 20 reactors have been placed in Safstor, with about a dozen being fully disassembled.

Most parts of a nuclear power plant do not become radioactive or are contaminated to very low levels, which allows the metal to be recycled.

The first US reactor that provided energy for the public at Shippingport, Pennsylvania (1957–1982) was decommissioned as a demonstration for the safe and cost-effective dismantling of a commercial-scale reactor. The defueling was completed in two years and five years later the site was released for other use without any restrictions. The decommissioning of the Shippingport reactor began in 1982 after about 25 years of operations. It was completely dismantled with the removal of the 900 ton reactor vessel used as a signal unit. The vessel was filled with lightweight concrete and placed on a specially designed barge. The barge travelled through 8,100 miles of a water route to its destination at the Hanford Washington State low level radiation site where it was buried. The decommissioning took over four years and proved that complete dismantlement of a commercial nuclear reactor was possible and cost effective.

Proven techniques and equipment are available to dismantle nuclear facilities safely and these have been demonstrated in the United States and elsewhere. The Shoreham reactor, which was shut down in 1989, without ever achieving authorization to operate at full power, was completely deactivated in 1994. In contrast, California's San Onofre 1, which was closed in 1992 after 24 years of operations, will not be fully decommissioned until about 2030.

Further Reading

"Beir VII: Health Risks from Exposure to Low Levels of Ionizing Radiation," *National Academy of Sciences*, 2006. https://nap.nationalacademies.org/resource/11340/beir_vii_final.pdf
A report that is part of ongoing studies on the health effects of low levels of ionizing radiation that are conducted by the National Academy of Sciences.
"Destruction of Long-lived Radioactive Waste," *Nuclear Energy Agency* (last modified February 3, 2012). https://www.oecd-nea.org/trw/intro/ens.html
An analysis of the nature of long-lived nuclear energy waste and its management.
Matthew L. Wald, "The Boring Truth About Nuclear Waste," *The Breakthrough Institute*, November 29, 2022. https://thebreakthrough.org/issues/energy/the-boring-truth-about-nuclear-waste#:~:text=When%20you%20think%20about%20nuclear,it%20isn%27t%20even%20waste
An analysis of misconceptions about nuclear waste by veteran science writer Matthew L. Ward.
"Radiation and Health Effects," *World Nuclear Association* (last modified April 29, 2024). https://world-nuclear.org/information-library/safety-and-security/radiation-and-health/radiation-and-health-effects.aspx
A discussion of radiation related to nuclear reactors.
Robert Peter Gale, MD, PhD, and Eric Lax, *Radiation: What It Is, What You Need to Know*, Vintage Books Random House, 2013.
A comprehensive tutorial on the nature of radiation, its use in medicine and its impact on health.
"TRISO Particles: The Most Robust Nuclear Fuel on Earth," *US Department of Energy, Office of Nuclear Energy*, July 19, 2019. https://www.energy.gov/ne/articles/triso-particles-most-robust-nuclear-fuel-earth
An overview of the robust properties of TRISO fuel particles.

17

Generation-IV Reactors

Nuclear is ideal for dealing with climate change, because it is the only carbon-free, scalable energy source that's available 24 hours a day. The problems with today's reactors, such as the risk of accidents, can be solved through innovation.

– Bill Gates[1]

Generation-IV Reactors: Restructuring, Innovating, Reengineering

The story of the development of advanced nuclear reactors is a remarkable tale of international cooperation in which more than a dozen countries engaged in a 30-year program of shared research and development. While collaboration among independent nations in the advancement of pure science has precedent, collaboration for the advancement of applied technology has never been witnessed on such a large scale. The joint pursuit of nuclear reactor innovation emerged from modest informal discussions into a unique model for development of one of the world's most consequential advanced engineering artifacts. Not only is this international initiative unique as a united quest for new technology, but also it is a joint effort involving countries that are often competitors rather than collaborators. In addition, this quest is one that began with a 30-year timeline which appears to be an achievable goal!

The impetus for international collaboration in science came from the devastation experienced by European countries during the Second World War. Not only was there a need for society in general to rebuild, but the European scientific community had experienced the exodus of many of its members to the United States. This depletion of human resources was particularly evident in the field of nuclear physics, which played such an important role at the conclusion of the war in 1945. Furthermore, study on the forefront of nuclear physics required costly equipment such as "atom smashers." Following the war, the United States institutionalized the national laboratory system that proved so successful in the Manhattan Project. In addition to Oak Ridge National Laboratory in Tennessee and Argonne National Laboratory in Illinois, the United States established

---

[1] Bill Gates, "Year in Review Letter," December 29, 2018. © Gates Notes, LLC.

Brookhaven National Laboratory in Long Island, NY in 1947. Brookhaven became a center for elementary particle physics with the decision in 1948 to build the Cosmotron, which was then the world's most energetic atom smasher.

CERN

As European scientists assessed the dominance in nuclear physics research in the United States in the late 1940s, they saw that to compete, that collaboration would be essential. In Paris in December 1951 at an intergovernmental meeting of UNESCO the first resolution for the establishment of a European Council for Nuclear Research (Conseil européen pour la recherche nucléaire, or CERN) was adopted. Twelve European countries ratified CERN in 1952, and its founding members were Belgium, Denmark, France, the Federal Republic of Germany, Greece, Italy, the Netherlands, Norway, Sweden, Switzerland, the United Kingdom, and Yugoslavia.

Based in Geneva, Switzerland, CERN became the largest elementary particle physics laboratory in the world with 23 member states, plus Israel. It has also been a center for innovation in the field of computer science.

Euratom

While the European Union was not established until 1993, the countries of Europe recognized the need for CERN-like collaboration in nuclear energy and nuclear physics. The European Atomic Energy Commission (EAEC or Euratom) was established in 1957. While the Euratom organization today is the same as that of the European Union, it is an independent organization and does not fall under the control of the European Parliament. Switzerland and the United Kingdom, while not members of the European Union, are associate members of Euratom.

Euratom was established about the same time as the development of civilian nuclear reactors to provide secure access to nuclear materials and technologies that facilitate peaceful use of atomic energy. Two key components of the single market established by Euratom are the free movement of capital to invest in the development of nuclear power and the free movement of experts to work at nuclear facilities. It also facilitates the highly regulated movement of nuclear goods, especially isotopes used in nuclear medicine. Euratom also coordinates the establishment of standards and regulations for the safe and secure handling and use of nuclear materials. While pursuing these important areas of nuclear collaboration, Euratom was not established, nor did it pursue active programs for nuclear technology innovation.

Generation-IV International Forum (GIF)

The Chernobyl catastrophe in 1986 brought nuclear energy development worldwide to a near halt for over a decade. The public was traumatized. Both the nuclear private sector and the related government agencies were paralyzed. It took some time for initiatives to emerge in an environment that recognized that new approaches to nuclear energy were needed. Incremental change was initiated in the 1990s with the development of safer, streamlined pressurized-water reactors (PWRs) and boiling-water reactors (BWRs), but forces were at work that called for a break with the past and the initiation of quite different designs.

With the successful precedents for international collaboration in pursuit of science and technology that were established by CERN and Euratom, a new collaborative effort in pursuit of innovative reactor technology emerged. An historic meeting of international participants was convened by the US Department of Energy in January 2000 to consider development of what they referred to as Generation-IV nuclear reactors. The initial group of nine countries included the United States, Argentina, Brazil, Canada, France, Japan, the Republic of Korea, the Republic of South Africa, and the United Kingdom. This working group discussed a framework that would support an ambitious agenda that resulted in a charter formalized in July 2001. Subsequently, five additional members joined the Forum: Switzerland in 2002, Euratom in 2003, the People's Republic of China in 2006, the Russian Federation in 2006, and Australia in 2016.

The GIF Charter noted the advantages of collaboration on research and development of advanced nuclear energy systems. The Charter prioritizes the development of competitively priced supply of energy while addressing nuclear safety, waste, proliferation resistance, and public perception concerns of citizens in countries in which the new technologies would be deployed.

A key role for GIF was the identification of priority areas for collaborative research and development. While in the early days of nuclear energy development various alternative designs were explored, the overwhelming emphasis quickly became the PWR and the BWR designs. That narrow focus was the result of the efforts of Admiral Hyman Rickover, who shaped much of the early work through his oversight of the successful nuclear navy and the desire of nuclear weapons countries to ensure a supply of plutonium for bombs that came as a by-product of the uranium-fueled water reactors.

GIF was structured to promote and encourage innovative nuclear reactor technologies and to bring together engineers and planners from multiple countries in carefully planned research agendas. Collaboration on specific projects was encouraged with members expected to be active in at least one significant collaborative project.

These collaborative efforts were structured in the context of a Roadmap formulated by a Roadmap Integration Team. In this initial activity of GIF, a team of over 100 technical experts contributed to the development of the Roadmap.

During 2001 and 2002, these experts considered 130 possible nuclear reactor designs for inclusion in the GIF agenda. This multiplicity arose from the fact that reactors could be built with different fuels, different coolants, different engagements of moderators, and different applications objectives. By the end of 2002, six basic systems were identified for development by GIF:

1. Gas-Cooled Fast Reactor System
2. Lead-Cooled Fast Reactor System
3. Molten-Salt Reactor System

4. Sodium-Cooled Fast Reactor System
5. SuperCritical Water-cooled Reactor System (SCWR)
6. Very High-Temperature Reactor System.

Only one of these systems involved the use of water: the SuperCritical Water-cooled Reactor (SCWR) System. When encountering water in day-to-day experience we expect to find the water to be in one of three states: solid (ice), liquid, or gas (steam). However, when there is sufficient pressure (more than 218 times the standard atmospheric pressure) and a high enough temperature (more than 705 °F/374 °C), water manifests itself in a fourth state in which the gas and liquid are indistinguishable from each other. This fourth state has properties that are particularly supportive of heat transfer. While desirable for a power reactor, the engineering challenges of building such a reactor are formidable and not likely to be solved soon. Therefore, we do not include a detailed analysis of that development path in this book.

While the establishment of GIF was a result of a US initiative, its administrative home became the Nuclear Energy Agency (NEA) of the Organization for Economic Cooperation and Development (OECD) located in Paris. William D. Magwood IV served as Director of the US Department of Energy Civilian Nuclear Program from 1998 to 2005. During that time, Magwood founded the GIF and was its first Director. Today, the OECD NEA, for which he serves as Director-General, is the Secretariat for the GIF.

Given that nuclear technology is promoted as a national enterprise for political and economic advantages and that, within nations, companies pursue that technology as proprietary initiatives, it is striking that the GIF Charter called for open research. It stated, "To the extent possible, the R&D fostered by the GIF should be open and non-proprietary."

It is astonishing that the RoadMap developed in 2002, with plans for reactor designs with a goal of implementation by 2030, appears to be meeting that schedule. During that almost 30-year period, the need for safe, economically viable alternatives to fossil fuel has become increasingly urgent. It appears increasingly likely that Generation-IV reactors will play a major role to help reduce global warming.

Generation-IV Goals

The GIF was established to develop reactor technology that met eight central goals. These goals are in four areas: sustainability, safety, economics, and protection against proliferation and terrorism.

They were formulated as follows:

1. Generation-IV nuclear energy systems will provide sustainable energy generation that meets clean air objectives and provides long term availability of systems and effective fuel utilization for worldwide energy production.

2. Generation-IV nuclear energy systems will minimize and manage their nuclear waste and notably reduce the long-term stewardship burden, thereby improving protection for public health and the environment.

3. Generation-IV nuclear energy systems will have a clear life-cycle cost advantage over other systems.

4. Generation-IV nuclear energy systems will have a level of financial risk comparable to other energy projects.

5. Generation-IV nuclear energy systems operations will excel in safety and reliability.

6. Generation-IV nuclear energy systems will have very low likelihood and degree of reactor core damage.

7. Generation-IV nuclear energy systems will eliminate the need for off-site emergency response.

8. Generation-IV nuclear energy systems will increase the assurance that they are very unattractive and the least desirable route for diversion or theft of weapons-usable materials and provide increased physical protection against acts of terrorism.

In summary, Generation-IV nuclear energy systems should be fail-safe and cost effective while managing waste and not being targets for terrorism or sources for proliferation. Here, fail-safe means that responses to accidents should rely only on phenomena that do not require operator intervention or specialized mechanisms.

Generation-IV Safety Feature: Not Water Cooled

Given that we are excluding the SCWR for this discussion, none of the other Generation-IV designs use water as a coolant. Eliminating water from the design is a significant safety feature since water introduces multiple vulnerabilities. At Chernobyl the large explosion that ripped apart the reactor was due to steam from coolant water that had overheated. Hence, the removal of water from the design eliminates the possibility of a steam explosion. Water was also the source of the Fukushima reactors' destruction, where overheated water oxidized the metal fuel containment units, which released hydrogen gas that accumulated and exploded in three of the Fukushima reactors. Hence, the removal of water from the design eliminates the possibility of production of dangerous hydrogen gas in the event of the reactor overheating.

In addition, the use of water as a coolant requires the system to be kept under high pressure since water boils at 212 °F (100 °C) and effective heat transfer requires the coolant fluid to be at a higher temperature. For water to attain a temperature of 617 °F (325 °C) without turning to steam, it must be subject to a pressure that is 150 times greater than atmospheric pressure. This requirement for keeping water in water-cooled reactors under high pressure is a challenging requirement that requires construction of specialized pressure vessels and use of pressurizers that add cost to reactor construction while creating vulnerabilities for damage and accidents. The basic designs that we are considering for Generation-IV systems all operate at or near atmospheric pressure.

Generation-IV Safety Feature: TRISO Fuel

A major design advance that can be used in all reactors that use U-235 fuel activated by thermal or slow neutrons is fuel that has multiple safety features known as TRISO, as discussed in Chapter 16.

Generation-IV Safety Feature—Fail-safe Designs

There are several fail-safe passive safe interventions that can automatically shut down reactors used in Generation-IV designs.

One of the most favored approaches is to use a freeze plug, which melts at high temperature and allows the contents of a molten-salt reactor to empty from its containment vessel into a large vessel in which a chain reaction cannot take place. The freeze plug is fabricated from lithium fluoride salts with melting temperatures in the range of 2192 °F (1200 °C). The first freeze plugs were developed at Oak Ridge National Laboratory in the early 1960s during the development of the first molten-salt reactors. Subsequently, detailed research on freeze plug design has been carried out in the United States, Europe, and China.

A passive safety feature demonstrated with liquid sodium-cooled reactors is the effect of thermal expansion on the fission process. In 1986 at Argonne National Laboratories safety demonstrations were conducted with the Integral Fast Reactor cooled by liquid sodium. One demonstration simulated the Three Mile Island accident, and another simulated the Fukushima accident. In both cases the reactor core experienced serious overheating, but due to the liquid-sodium design the fuel expanded and prevented neutrons from sustaining a chain reaction. The reactor shut itself down without any human or technological intervention. The shutdown was entirely the result of the behavior of natural processes.

A third passive safety feature used in gas-cooled reactors is automatic insertion of fuel rods due to overheating. In this design, the fuel rods are held by natural magnets that can lose their magnetic properties at a specific temperature. This temperature is known as the Curie point and is different for different magnetic materials. The magnets in the reactor are selected based on the temperature chosen for automatic shutdown.

Another passive safety feedback mechanism results from bubbles or voids that might form during heating of a molten salt reactor. These voids reduce the neutron flux that is causing heat generation by reducing the density of the reactor's fuel. Hence, overheating that causes boiling automatically results in lowering the reactor temperature. The opposite contributed to the accident at Chernobyl. The Russian-designed RBMK reactor had a design flaw of a positive void coefficient. In that case, as the temperature of the reactor increased and caused bubbling of the water coolant, the absence of water, which was acting as a moderator for the neutrons, allowed more neutron-induced reactions to take place with a resulting increase in the reactor's temperature. This process contributed to the Chernobyl steam explosion. Since molten salt is a carrier of fuel (rather than a moderator in these reactors) voids reduce, rather than enhance, energy production.

The physics of nuclear absorption provides yet another fail-safe mechanism. As the temperature of the fuel increases in the event of overheating, the motion of the U238 nucleus allows a higher probability for neutron absorption, thus reducing the chances for absorption by U-235 that is driving the increase in temperature. This mechanism is known as Doppler Broadening and is a well-known property of U-238.

Further Reading

"History and Achievements of GIF," *Gen IV International Forum*. https://www.gen-4.org/gif/jcms/c_9334/origins

Comprehensive information regarding the development and status of Generation-IV nuclear reactors is found at the website of the Generation IV International Forum. Of particular interest is their collection of annual reports and their informative webinars, both of which can be accessed from the website. The secretariat for the Forum is the Nuclear Energy Agency (NEA) that is associated with the Organization for Economic Cooperation and Development (OECD).

Thomas Schulenberg, *The Fourth Generation of Nuclear Reactors: Fundamentals, Types and Benefits Explained*, Springer, 2022.

This book is aimed at the non-specialist. It presents the fundamentals of Generation-IV reactor designs and reports on their significance for the economy and society.

18

Molten-salt Reactors

Our reactors don't just produce electricity. We are designing an entire ecosystem that takes advantage of the benefits of a thorium molten-salt reactor future.

— Flibe Energy, Inc.[1]

Molten-salt Reactor: Overview

While the designation for Generation-IV designs in the RoadMap was not formalized until 2002, several of the most promising designs had been pursued from the earliest days of nuclear energy. This is particularly true of the molten-salt reactor (MSR) that was studied at Oak Ridge National Laboratory.

The MSR history is closely linked to the work of Alvin Weinberg, who received a doctorate in mathematical biophysics from the University of Chicago in 1939 and joined the Metallurgical Laboratory division of the Manhattan Project in Chicago in 1941. The focus of the Metallurgical Laboratory was production of plutonium for bombs using nuclear reactors. Reactors containing uranium-238 (U-238) absorbed neutrons thereby transforming into plutonium-239 (Pu-239).

At the Metallurgical Laboratory, Weinberg was enlisted into a theoretical physics group that engaged in calculations on neutron capture by uranium. The work of this group was taken over by DuPont Corporation, which oversaw the development of the first reactors at the Hanford, WA site. As the work at Hanford proceeded under the management of DuPont, the Metallurgical Laboratory theoretical group turned its attention to other possible designs, for which many possibilities existed.

This group, under the direction of Eugene Wigner, realized that in addition to U-235 and Pu-239 there was a third isotope that could sustain a chain reaction. That is U-233 that does not exist in nature, but which can be produced by bombarding the element thorium with neutrons. Thorium is an element with 90 neutrons and its most common isotope is thorium-232 that has 142 neutrons. When thorium-232 absorbs a neutron it transforms into the element protactinium-233 with a half-life 22 minutes. This happens because of natural radioactive decay in which a neutron emits an electron and transforms into a proton. A second such transformation then changes protactinium-233 into U-233.

---

[1] Flibe Energy, Inc., 2025. https://flibe.com/technology, accessed January 30, 2025.

Nuclear Energy. Edward A. Friedman, Oxford University Press. © Edward A. Friedman (2025). DOI: 10.1093/9780198925811.003.0018

The use of thorium-232 to produce U-233 as a fuel for nuclear reactors is an attractive prospect, since it cannot easily be the source of plutonium-based nuclear weapons and since the absence of U-238 in the reactor eliminates the source of the long-lasting radioactive isotopes produced when U-238 absorbs neutrons. Weinberg, who gained this understanding of the potential for using thorium as a reactor fuel during the Manhattan Project, later used this knowledge to implement the first thorium reactor in his later programs.

At the end of the Second World War, Wigner's theoretical physics group moved to Oak Ridge National Laboratory where Weinberg became director of research in 1948. In the competition among the services, the Navy was active in post-war development of a nuclear submarine, which led to the Army establishing a program for a nuclear-powered bomber. While the possibility of this effort succeeding was always seen as a long shot, it did promote fruitful research and development activities at Oak Ridge. While Weinberg realized there could be a thousand distinct reactor designs given the multiplicity of possible fuels, coolants, moderators, and control devices, he favored the MSR design using thorium or uranium as the fuel.

Under Weinberg's direction the world's first molten-salt-fueled and -cooled reactor became critical in 1954. This aircraft nuclear propulsion reactor set a record high temperature at 1,600 °F (870 °C). However, radiation dangers to the crew of the aircraft were seen as unmanageable and the potential for contamination on the ground in the event of a crash added to the loss of interest in this program. It was also the case that aerial refueling became routine, thus allowing long-distance flights without nuclear energy. Weinberg became director of Oak Ridge in 1955 and witnessed the cancellation of the Aircraft Nuclear Propulsion program in 1961.

Weinberg then oversaw the transition of the Aircraft Nuclear Propulsion project to a general research program on MSRs known as the MSR Experiment. Construction of the test reactor for this program was initiated in 1962 and went critical in 1965. It was a 7.4-megawatt (MW) thermal reactor that produced heat. It was not connected to generators to produce electricity. The reactor operated for five years, during which time it was operational 80% of the time. A wide range of reactor properties were studied including adding fuel and removing spent fuel from the molten salt during operations, along with tests of the stability of the materials used during long-term operations. The overall assessment of the studies proved the viability of the design. During 1964, the final year of the program, sufficient U-233 produced from thorium was available. It was successfully used as a reactor fuel.

Oak Ridge, under Weinberg, was the primary location for research and development of the MSR design. Weinberg was a tireless advocate for this proven technology. However, he was not as effective in influencing oversight agencies and members of Congress as were others, most prominently Admiral Hyman Rickover, whose views held sway in the White House, Congress, and the Atomic Energy Commission (AEC). In 1973, Weinberg was fired as director of Oak Ridge for not endorsing the Nixon administration's reactor development program. His departure from that position resulted in an interruption in MSR development.

The obvious advantages of the MSR design attracted enthusiastic supporters of the concept and its ability to use thorium as a fuel made it particularly appealing. In 2011, five years after Weinberg's passing, environmentalists in the United Kingdom initiated the Alvin Weinberg Foundation to promote the use of thorium as a reactor fuel.

This development attracted a great deal of attention due to the participation of Bryony Worthington Baroness Worthington as a founder. She was the youngest woman serving in the House of Lords. The Foundation succeeded in promoting Weinberg's work during six years of its activity until it was dissolved in 2017. Even without the Foundation as a driving force, there exists considerable support for the development of a thorium-fueled MSR with an almost cult-like cadre of advocates.

Given the extensive research that was pursued on MSRs at Oak Ridge under the guidance of Alvin Weinberg and the many safety features associated with these designs, it is unsurprising that in the review of potential Generation-IV nuclear reactors that MSRs were identified for further development.

There are many diverse types of MSRs that are possible with a clear dividing line between reactors that use slow neutrons to sustain fission reactions with graphite moderating materials used to slow the neutrons and reactors that use fast neutrons to activate fission in plutonium that is bred from U-238.

Since 2002, there have been several significant development initiatives centered on the use of the MSR design. The next sections discuss some examples.

Terrestrial Energy

Terrestrial Energy is a Canadian based company that is pursuing a small modular reactor design known as the integral molten-salt reactor (IMSR). Given that the regulatory regime in Canada is more interactive with applicants than in the United States, there is a clear advantage to development of a new reactor design in Canada. Terrestrial Energy is designing the ISMR for applications that use high-temperature output heat for industrial processes. As a small modular reactor, Terrestrial Energy looks forward to large-scale factory production of reactors.

China Wuwei Reactor

China is pursuing an aggressive program of MSR development that uses thorium. A prototype 2-MW demonstration reactor using a mixture of thorium and uranium fuel components is proceeding in anticipation of construction of larger reactors employing this design. These will be the first operational MSRs since the shutdown of the Oak Ridge program in 1969. While China is incorporating knowledge gained in the Oak Ridge program, the design will be fully owned by China, which hopes to employ these reactors in country, as well as in their export market.

ThorCon

ThorCon is an American company promoting use of a MSR design to be constructed as a floating power plant on a barge. The reactor's fabrication is planned for assembly line production in a shipyard with delivery of the power plant to any major waterway shoreline. The reactors would be delivered as a sealed unit and never opened on site. All reactor maintenance and fuel processing would be done at an offsite location. The power station is planned as two 250-MW electric units to be replaced every four years. Thorcon is pursuing implementation of this plan in Indonesia in collaboration with Indonesian partners. The conceptualization of an MSR built on a barge is not unique to ThorCon. Denmark's Seaborg Technologies is engaged in similar plans, as are other companies as well as the national program in Russia.

Flibe Energy

Flibe Energy was founded by Kirk Sorensen who has been an outspoken advocate for the use of thorium as a nuclear fuel. Sorensen worked on reactor development for NASA planning nuclear energy for use on the Moon when he realized the advantages of a thorium strategy for Earth. The name Flibe is derived from the shorthand notation identifying the salts that are used in the Flibe reactor (lithium fluoride (LiF) and beryllium fluoride (BeF_2). These salts accommodate the fuel at atmospheric pressure and at operating temperatures of 1112–1292 °F (600–700 °C).

The solution is chemically stable and supports the transformation of thorium into U-233 that undergoes fission. After the reactor operations are started with the insertion of pre-developed U-233, the ongoing energy generation is self-sustaining. Flibe Energy is carrying the aspirations of Alvin Weinberg forward but has not yet succeeded in implementing a commercially viable reactor. However, this remains a promising avenue for development.

Moltex

Moltex is a Canadian company that is under contract to build an MSR that is specifically designed to use nuclear waste as a fuel. Moltex anticipates developing an operational MSR in the early 2030s.

Kairos

Kairos Power, based in Albuquerque, New Mexico is developing a fluoride salt-cooled high-temperature reactor that uses TRISO fuel. In 2020, Kairos Power was selected by the US Department of Energy to receive $303 million over a seven-year period to support work to build a demonstration reactor at Oak Ridge National Laboratory.

Further Reading

"Molten Salt Reactors," *World Nuclear Association* (last modified September 10, 2024). https://world-nuclear.org/information-library/current-and-future-generation/molten-salt-reactors.aspx
A history and status report on molten salt reactors.
"Molten Salt Reactors," *Wikipedia* (last modified March 9, 2025). https://en.wikipedia.org/wiki/Molten-salt_reactor
A report on molten salt reactors.
Richard Martin, *SuperFuel: Thorium, the Green Energy Source for the Future*, Springer, 2012.
Robert Hargraves, Thorium: Energy Cheaper than Coal, Creative Space Independent Publishing Platform, 2012.
Both books present advocacy positions for the potential for thorium, molten-salt reactor technology.

19

Liquid-sodium Reactors (LSRs)

Today, we broke ground on the first-ever natrium plant in Kemmerer, Wyoming. This next-generation nuclear power plant is a big step towards safe, abundant, zero-carbon energy.

— Bill Gates on X, June 11, 2024

Sodium-Cooled Fast Reactors: Overview

Sodium-cooled fast reactors (SFRs) use liquid sodium as a coolant instead of water. SFRs can be more efficient and produce less waste than traditional water-cooled reactors. These reactors use fast neutrons to breed fuel by transforming uranium-238 (U-238) into plutonium-239 (Pu-239), one of the three isotopes that can sustain a chain reaction—along with uranium-233 (U-233) and uranium-235 (U-235), which is then consumed in heat production. These reactors operate at a higher temperature and lower pressure than water-cooled reactors. The higher temperature sustains more efficient heat transfer, and the lower pressure allows greater safety through the elimination of pressurizing equipment.

Russia has pioneered in the use of SFRs and is investing in its development. Other countries with SFR programs include India, Japan, France, Canada, South Korea, and the United States.

Russia

SFRs have been used in Russia since the 1970s, and the country has a long history of operating this type of reactor. The first commercial SFR, known as the BN-350, began operation in 1973 in Aktau, Kazakhstan. The reactor was used primarily for electricity generation but also provided heat and desalinated seawater for the local community.

In the 1980s, Russia began operating the BN-600, a larger and more advanced SFR still in operation today. The BN-600 was designed to generate electricity and to serve as a prototype for future fast reactors. In the 1990s, Russia began developing a next-generation SFR known as the BN-800, which began operating in 2016 and is currently the largest SFR in the world. It produces nearly 800 megawatts (MW) of electrical

Nuclear Energy. Edward A. Friedman, Oxford University Press. © Edward A. Friedman (2025). DOI: 10.1093/9780198925811.003.0019

power. Russia has also been involved in international efforts to develop SFRs, including collaborations with France, Japan, and China.

There has been reluctance to pursue development of SFRs due to the highly reactive nature of liquid sodium, which reacts violently with water and air, leading to the release of hydrogen and the possibility of explosions. There have been several accidents involving SFRs in Russia. The most notable incident occurred in 1986 at the BN-350 reactor in Kazakhstan, where a sodium leak resulted in a fire and radiation release. There have also been incidents at the BN-600 and BN-800 reactors, although these were less severe. Despite these incidents, Russia remains committed to the development and operation of SFRs. These vulnerabilities require careful attention to safety measures.

SFRs in the United States

The possibility of an SFR was of interest to developers in the United States from the earliest days of nuclear reactor activity. In the early 1950s the Fermi 1 reactor was designed and built in Monroe, Michigan, by the Atomic Energy Commission (AEC). It had an electric power output of 20 MW.

The Fermi 1 reactor began operation in August 1963 and, in October of the same year, was shut down after a partial meltdown caused by a blockage in the reactor's cooling system. The block led to a loss of coolant flow and subsequent overheating of the fuel. This caused some of the fuel to melt and release radioactive gases into the containment building.

After the partial meltdown, the reactor was permanently shut down and decommissioned. The cleanup of the site took several years and cost millions of dollars. The incident at Fermi 1 led to significant changes in the design and regulation of nuclear power plants in the United States, including the establishment of the Nuclear Regulatory Commission (NRC).

Terrapower—Traveling-wave Reactor

The TerraPower traveling-wave reactor (TWR) is a type of nuclear reactor design that operates on a different principle than traditional nuclear reactors. This design s that of a breeder reactor structured to immediately burn the fuel created. The TWR is conceptualized as a "traveling wave," in which the interface at which the fuel is created and then consumed moves through the system like a wave.

The TWR uses depleted uranium as the basic fuel. Depleted uranium is extracted from the waste of water-cooled reactors that have used the U-235 isotope for fission. The depleted uranium is formed into rods, which are placed in the TWR core, where there are rods with enriched U-235 used to start the process by initiating a chain reaction in rods of depleted uranium surrounded with it. Neutrons from that chain reaction are absorbed by U-238, leading to Pu-239 production. That Pu-239 is immediately engaged in the chain-reaction process. Periodically, to sustain the fission reactions a fuel-handling machine shuffles the fuel rods, swapping the expired fuel at the center of the core for fresh fuel rods from the outer edges. This enables the reactor to operate for decades without refueling, providing continuous power output. This processing pattern was determined using extensive computer modeling with supercomputers.

This remarkable design was first conceptualized in 1958 by Soviet physicist Savely Moiseevich Feinberg, who called it a "breed and burn" reactor. The concept was further developed in 1995 by Edward Teller and the prolific inventor Lowell Wood. Bill Gates was then intrigued by the notion of a long-running self-contained reactor and devoted a considerable investment of supercomputer modeling to actualize the modeling with a demonstration reactor. Gates entered into an agreement with the China National Nuclear Corporation (CNNC) for joint development of the TWR in 2015. However, the program was abandoned in 2019 when the Trump administration constrained technology agreements between the United States and China.

Given that TerraPower, which is owned by Bill Gates, holds the intellectual property rights for this revolutionary design, it is likely that they will move forward with its development prior to 2030.

France: The Advanced Sodium Technological Reactor for Industrial Demonstration (ASTRID)

The Advanced Sodium Technological Reactor for Industrial Demonstration (ASTRID) is a fourth-generation SFR designed by the French Alternative Energies and Atomic Energy Commission (CEA).

The ASTRID reactor was designed to be a next-generation nuclear reactor that could provide sustainable and safe nuclear energy with improved performance and lower waste production compared to previous generations of reactors. The reactor was also designed to be a stepping stone toward commercialization of fast reactors.

The design of the ASTRID reactor features a compact core and an innovative fuel design, which uses a combination of uranium and plutonium oxide fuels to improve fuel use and reduce waste production. The reactor also incorporates passive safety features, such as a natural circulation cooling system, to ensure safe operation even in the event of a loss of power or other emergencies.

Construction of the ASTRID reactor began in 2014, but in 2019 the French government announced that it was canceling the project due to budget constraints. As a result, the ASTRID project was halted before the reactor was completed.

The ASTRID reactor was part of a larger effort by the French nuclear industry to develop advanced reactor technologies, which would help to secure France's position as a leader in nuclear power and maintain its nuclear energy supply chain. Despite the cancellation of the ASTRID project, the French nuclear industry continues to invest in the development of advanced nuclear technologies, such as small modular reactors (SMRs) and other Generation-IV reactors.

India: The Prototype Fast Breeder Reactor (PFBR)

Located in Kalpakkam, India, the Prototype Fast Breeder Reactor (PFBR) is a nuclear reactor designed to generate electricity using a fast breeder reactor (FBR) design.

The PFBR was designed by the Indira Gandhi Centre for Atomic Research (IGCAR). It is a 500-MW electric FBR. The reactor uses liquid sodium as a coolant and mixed oxide (MOX) fuel, which is a blend of plutonium and uranium oxides.

The FBR design uses fast neutrons to breed more fuel in the reactor than it consumes, resulting in a self-sustaining chain reaction. This design can generate large amounts of energy from a relatively small amount of fuel.

Construction of the PFBR began in 2004, and it was initially scheduled to be completed by 2010. However, the project faced several delays due to technical issues and safety concerns. Then, in 2011, the reactor achieved its first criticality, when a self-sustaining chain reaction was established in the reactor. However, it has yet to be fully operational due to further technical issues and safety concerns.

Prime Minister Narendra Modi was in Kalpakkam on March 4, 2024, for the initiation of the first core loading. This marked a significant step in India's nuclear development planning. The Kalpakkam PFBR will eventually be used with thorium as a blanket for the generation of U-233 fuel. India's long-term goal is to use its large reserves of thorium in an advanced reactor strategy for energy production.

Japan: The Monju Reactor

The Monju reactor was an FBR located in Tsuruga, Fukui Prefecture, Japan. It was designed to generate electricity by using the fast-neutron spectrum to produce more fissile material than it consumes.

The construction of the Monju reactor began in 1985, and it achieved criticality in April 1994. However, a fire broke out during a sodium leak in December of the same year and damaged some of the reactor's components. The reactor remained closed for over a decade for repairs.

The reactor resumed operation in May 2010, but a few months later, it was shut down again due to a fuel exchange device falling into the reactor vessel during a maintenance check. Further, it was discovered that the operator had falsified safety records, leading to the resignation of the head of the Japan Atomic Energy Agency (JAEA).

After this troubled history and numerous accidents, the Japanese government decided to decommission the Monju reactor in December 2016. The decommissioning process is expected to take around 30 years, and it will cost billions of dollars.

Canada: The Advanced Reactor Concepts (ARC)

The ARC-100, a fourth-generation SMR design, is based on the Integral Fast Reactor (IFR) developed at Argonne National Laboratory. The Advanced Reactor Concepts (ARC) reactor design and IFR are related in that they have similar fast sodium-cooled designs. The IFR was developed in the 1980s using a prototype reactor that was subject to exacting safety trials. When loss of coolant conditions were imposed that paralleled the accidents at Three Mile Island and at Fukushima, the IFR automatically shut itself down. This was a result of the expansion of the liquid fuel that disabled the continuing chain reaction process. The IFR also succeeded in demonstrating the recycling of the spent fuel through pyroprocessing (see South Korea). Despite the success of this reactor, the political environment in the United States in the 1990s was hostile to nuclear energy and the Clinton Administration cut off funding and closed the program. At that time, John Kerry led the opposition to the IFR in the Senate.

The ARC-100 has scaled the energy output of the IFR up to 100 MW from the 20-MW output of the IFC reactor. The ARC-100 also has a unique fuel form factor that enables recycled nuclear fuel. This fuel design avoids the need for fuel reprocessing while still achieving high levels of fuel use and safety.

Implementation of this SMR is under development for use at the Point Lepreau Nuclear Generating Station in New Brunswick, Canada in the early 2030s.

In summary, while the ARC reactor design and IFR share some similarities in terms of their use of liquid-sodium coolant and advanced fuel cycle concepts, they are distinct designs with different fuel forms and objectives.

Pyroprocessing

Used successfully at the Argonne National Laboratory IFR, pyroprocessing uses spent fuel, which is subjected to high temperatures to separate fissile uranium and plutonium from the other materials in the spent fuel. The spent fuel is then chopped up into small pieces and placed in a furnace with an inert gas at 1652 °F (900 °C). The heat breaks down the spent fuel into its component parts. This allows the separate extraction of uranium and plutonium. While plutonium is available for use as a reactor fuel, it creates a proliferation risk that must be addressed.

South Korea

While South Korea has had on and off support for indigenous use of nuclear energy, it has been the home of active reactor development for export. The Korea Advanced Liquid Metal Reactor, known as Kalimer, is an SFR being developed by the Korea Atomic Energy Research Institute (KAERI). It is a fourth-generation reactor design that uses liquid sodium as a coolant.

The design of Kalimer incorporates several advanced features, such as a compact core design, a passive safety system, and a fuel cycle that includes the use of pyroprocessing technology to recycle used fuel. These features are intended to enhance the safety, efficiency, and sustainability of the reactor.

The development of Kalimer began in the late 1990s, and a prototype reactor was constructed at KAERI's facility in Daejeon, South Korea. The reactor achieved first criticality in 2010 and was connected to the grid for the first time in 2013. The prototype reactor has a capacity of 80 MW thermal and is designed to operate for 30 years.

In addition to the prototype reactor, KAERI has also been working on the development of a larger-scale commercial version of Kalimer, with a capacity of 600 MW thermal. This reactor is under construction with an anticipated completion date in the early 2030s.

The Kalimer has been the subject of significant international cooperation, with KAERI collaborating with organizations such as the International Atomic Energy Agency (IAEA), the US Department of Energy, and the French Alternative Energies and Atomic Energy Commission on various aspects of the reactor's design and operation.

Germany

The German Deutsch–Französischer Versuchsreaktor (DFR) was a joint research project between Germany and France designed to test and develop new nuclear technologies. The reactor was in Karlsruhe, Germany, and was in operation from 1981 to 1988.

The DFR was an FBR, which means that it used a fast-neutron flux to produce more fuel than it consumed. This technology can greatly increase the efficiency of nuclear power generation and reduce the amount of nuclear waste produced. The reactor used liquid sodium as a coolant, which has a high heat capacity and can transfer heat quickly and efficiently.

The DFR project was initiated in 1964 and was jointly funded by Germany and France. The design of the reactor was completed in 1971, and construction began in 1973. The reactor achieved criticality for the first time in 1981 and operated for seven years before it was shut down in 1988 because of technical problems.

One of the main issues with the DFR was the potential for sodium leaks, which could cause fires and explosions. There were several incidents of small sodium leaks during the reactor's operation, but these were all contained without causing any severe damage. However, in 1988, a larger sodium leak occurred, which led to a fire and significant damage to the reactor. This incident led to its permanent shutdown.

Overall, the DFR project provided an important milestone in the development of nuclear technology, and the knowledge gained from the project has been used to improve the safety and efficiency of nuclear reactors. Further development of this reactor design in Germany is unlikely.

China: The Fast Sodium Reactor (FSR) Program

China's FSR program began in the 1960s and has gone through several design changes and operational phases over the years. China's first fast neutron reactor (FNR), the Chinese Experimental Fast Reactor (CEFR), was built in 1980 based on a French design. It had a capacity of 20 MW.

In the 1990s, China began designing its own fast neutron reactors, starting with the 65-MW CEFR, which was completed in 2010. This reactor used mixed oxide (MOX) fuel and operated for over ten years, accumulating a total of 65,000 hours (about 7.5 years) of operation.

China's next fast reactor design was the CFR-600, which began construction in 2011 and is still under development. China's pursuit of this technology is viewed with suspicion by many observers in Europe, Japan, and the United States since reactors with this design can be used both for civilian power production and for nuclear weapons development since weapons-grade plutonium can be separated from the spent fuel. While China has not been forthcoming regarding progress with this reactor, satellite photos taken in October 2023 indicate that considerable development has taken place at that site in Fujan.

China's FSR program has also been used to develop several other technologies, such as fuel fabrication, fuel reprocessing, and waste management. The program has also been used to develop several safety systems, such as emergency core cooling and containment systems.

Further Reading

Charles E. Till and Yoon Il Chang, *Plentiful Energy: The Story of the Integral Fast Reactor: The Complex History of a Simple Reactor Technology, with Emphasis on its Scientific Basis for Non-specialists,* Creative Space, 2011.
A report on the history of the sodium fast reactor research and development that took place at Argonne National Laboratory in the 1980s and early 1990s.
"Sodium-cooled Fast Reactor," *Wikipedia* (last modified December 3, 2024). https://en.wikipedia.org/wiki_Sodium-cooled_fast_reactor
A review of the history and current developments for sodium-cooled fast reactors.

20

Liquid Lead-cooled Fast Reactors

Lead has the highest atomic number of any stable element and three of its isotopes are endpoints of major nuclear decay chains of heavier elements.[1]

Lead-cooled Fast Reactors (LFRs): Overview

Lead-cooled nuclear reactors are an advanced nuclear reactor design that uses liquid lead or lead–bismuth eutectic (LBE) as a coolant instead of water. A eutectic mixture is a homogeneous blend of two or more constituent components that melts at a fixed temperature that is lower than the melting temperature of its components. Lead has one of the lowest melting points of any metal (621 °F (327 °C)). The melting temperature of lead–bismuth eutectic is much lower at 255 °F (124 °C).

Liquid-lead reactors share many of the advantages of liquid-sodium reactors while providing the feature that liquid lead is not nearly as chemically reactive as sodium, thus eliminating the dangers of explosions that occur when liquid sodium contacts air or water.

Liquid-lead-cooled reactors operate at high temperatures, allowing efficient heat transfer from the reactor's core to associated electric generators. Another advantage of lead coolant is its low rate of absorption of neutrons, which allows a chain reaction to activate with less highly enriched uranium (HEU) than other designs. Also, by operating at atmospheric pressure and being naturally self-cooling, this design enjoys important intrinsic safety features, removes the need for auxiliary pumps, and reduces the cost and complexity of the design.

Other advantages of lead as a coolant are the fact that it reflects neutrons, thus assisting in maintaining sufficient neutron flux to engage in a chain reaction. Lead also absorbs gamma rays and helps support a low level of radioactivity near the reactor.

Lead-cooled reactors also have the advantage that they can be designed to operate using a variety of nuclear fuels, including uranium, plutonium, and thorium. They can also be designed to operate in different configurations, such as a small modular reactor or a large-scale power plant.

---

[1] "Lead," *Wikipedia* (last modified November 2, 2024). https://en.wikipedia.org/w/index.php?title=Lead&oldid=1254922223, accessed November 11, 2024.

Nuclear Energy Edward A. Friedman, Oxford University Press. © Edward A Friedman (2025). DOI: 10.1093/9780198925811.003.0020

The disadvantages of lead as a coolant are that it does not conduct heat as well as sodium and that it can cause corrosion of the reactor components. Also, lead is toxic for humans, and it is not easily purged from the body.

Russia

Russia began development of a 300-megawatt (MW) electric lead-cooled reactor in 2020 known as the BREST-300. Plans to build a 1200-MW electric lead-cooled reactor have not yet been finalized. The BEST-300 will be the main facility of the pilot demonstration energy complex being constructed at Seversk, the Siberian Chemical Combine site with a target date for startup in 2026.

The early development of lead-cooled fast reactors (LFRs) began in the Soviet Union with a program to power nuclear submarines using that technology. The first Soviet nuclear-powered submarine reactor, the K-3 Leninsky Komsomol, used pressurized water technology and was operational in 1957. However, the Soviet Union also pursued research on LFRs for submarines, and by the 1970s had developed a small-scale experimental LFR for a research submarine. The first Project 705 Lira experimental lead-cooled nuclear reactor to propel a submarine was launched in 1971. Several vessels of this type were built with production ending in 1981.

Despite the potential advantages of LFRs, including their high-power density and the ability to use spent nuclear fuel as reactor coolant, the use of LFRs for submarines has not been widely adopted. This is partly due to concerns about the toxicity and corrosiveness of lead–bismuth coolant, as well as the potential for leaks and accidents. However, research on LFRs for submarines continues, with Russia and China among the countries exploring the technology.

The European Union ELSY Project and the Romanian Alfred Project

The European Lead-cooled System (ELSY) was initiated in 2006 by the European Commission to design and develop an LFR to generate electricity. This project is being pursued by a consortium of European research groups with active participation from Italy, the Netherlands, Belgium, and Germany.

The Advanced Lead-cooled Fast Reactor European Demonstrator (ALFRED) is an LFR currently being developed by the Romanian National Institute for Research and Development in Nuclear Engineering (INR), in collaboration with several other European countries. There is an overlap of the ALFRED team with that of ELSY.

The initial design concept for ALFRED was developed in the late 1990s, and since then, several design iterations have been made to improve its efficiency and safety features. The reactor will have a thermal power output of 300 MW and is designed to use lead as both the coolant and the neutron moderator.

ALFRED's primary mission is to demonstrate the feasibility of LFRs as a viable option for nuclear power generation. The reactor's design includes several innovative features, such as a passive decay heat removal system and a dual coolant system that uses both lead and helium. These features are intended to enhance the reactor's safety and reliability, making it less vulnerable to potential accidents or failures.

The operational history of ALFRED is still in the development stage, with plans to build a demonstration unit in Romania. The project has received funding from the European Union.

Sweden

The Swedish Advanced Lead Reactor (SEALER) is an LFR design developed by the Swedish company LeadCold Reactors AB to produce 3–10 MW of electricity. It uses lead as a coolant and a fuel consisting of uranium oxide dispersed in a graphite matrix. The SEALER also has a passive safety system and is designed to be transportable.

The SEALER is based on well-established technology and incorporates advanced safety features and modular design that makes it suitable for a wide range of applications. A prototype system is anticipated prior to 2030.

Belgium

The Multi-purpose Hybrid Research Reactor for High-tech Applications (MYRRHA) is a planned multi-purpose research reactor that is being developed by the Belgian Nuclear Research Centre (SCK CEN). The MYRRHA is intended to be a flexible and versatile research facility that can operate in a wide range of configurations, including a subcritical reactor, and a critical reactor.

A subcritical reactor is one that produces fission without having achieved criticality. The needed neutrons are provided by particles from an accelerator causing the ejection of neutrons from a heavy nucleus.

The design of MYRRHA is based on a liquid LBE-cooled and moderated system, which offers several advantages over traditional water-cooled reactors. These advantages include higher safety margins, better resistance to extreme temperatures, and the ability to operate in a subcritical configuration using a proton accelerator.

The MYRRHA project was initiated in 1998, and since then, it has undergone several design iterations and feasibility studies. In 2010, the Belgian government officially approved the project, and construction of the first phase of the reactor began in 2017. The first phase of the project, expected to be completed by 2026, involves the construction of a proton accelerator and a subcritical assembly for research and development.

The second phase of the MYRRHA project involves the construction of a full-scale, fully operational research reactor. This phase of the project is expected to be completed by 2035, and the reactor will be used for a wide range of research activities, including materials science, nuclear physics, and medical research.

The MYRRHA is considered an important project for the future of nuclear research in Europe and has received funding and support from several EU member states. MYRRHA is expected to play a key role in the development of next-generation nuclear technologies, and it is expected to help pave the way for a more sustainable and secure energy future.

United States

The Small, Sealed, Transportable, Autonomous Reactor (SSTAR) was a conceptual design for a compact nuclear reactor that was being developed at the Lawrence Livermore National Laboratory by the United States Department of Energy (DOE) Nuclear Energy Research Initiative (NERI) in the mid-2000s. The SSTAR reactor was intended to be a safe, reliable, and cost-effective source of nuclear power that could be used in a variety of applications, such as powering remote military bases, providing electricity to isolated communities, and supporting space exploration missions.

The SSTAR reactor was designed to be a small, modular reactor that could be transported and installed easily. The reactor would be sealed and autonomous, meaning that it would be self-contained and require no maintenance or refueling for many years of operation. The reactor was based on a liquid metal-cooled design, using an LBE coolant, and was intended to operate at a temperature of around 932–1112 °F (500–600 °C).

The SSTAR reactor was first proposed in 2002 as part of the DOE's NERI program aimed at developing advanced nuclear technologies. The design was refined over the next few years, with a focus on improving safety, reliability, and cost-effectiveness. The reactor was designed to be inherently safe, with passive cooling systems that would prevent overheating in an emergency. The design also incorporated advanced materials and manufacturing techniques to reduce costs and improve performance.

Despite its promising design, the SSTAR reactor was never built or operated. The DOE's funding for the project was limited, and the program was eventually canceled in 2009 due to budget constraints. However, the SSTAR design has influenced other advanced nuclear reactor designs, and some of its features have been incorporated into other projects, such as the integral molten-salt reactor (IMSR) and the molten-salt reactor experiment (MSRE).

Summary

While the United States does not have an active LFR program, efforts in Russia and Europe are actively moving to establish this technology as a competitor in the post-2030 drive to achieve a net-zero carbon world.

Further Reading

"Lead-cooled Fast Reactor," *Gen IV International Forum*. https://www.gen-4.org/generation-iv-criteria-and-technologies/lead-fast-reactors-lfr

"Lead-cooled Fast Reactor," *Wikipedia* (last modified March 11, 2025). https://en.wikipedia.org/wiki/Lead-cooled_fast_reactor

These two reports deal with the characteristics of and developmental status of lead-cooled nuclear reactors.

21

High-temperature Gas-cooled Reactors

The high operating temperature of high-temperature gas-cooled reactors potentially enable [non-electric] applications such as process heat or hydrogen production.[1]

High-temperature Gas-cooled Reactors (HTGRs): Overview

High-temperature gas-cooled reactors (HTGRs) using helium gas are attracting a great deal of attention due to their having multiple safety features and their potential of operating at high temperatures at 1472–1832 °F (800–1000 °C). At those temperatures, direct heat from the reactor can be used to produce hydrogen gas sought after for many energy applications and to sustain a variety of industrial processes with direct heat.

HTGRs can also be used with TRISO fuel that adds additional operational advantages and safety features. In the event of an accident, the helium gas stops circulating but natural air circulation along with radiation from the basic structural elements can carry off residual heat.

History

Before TRISO fuel development, in the early days of nuclear energy development, helium-cooled reactors were seen as having great potential. Plans were formulated in 1965 for the construction of a gas-cooled 330-megawatt (MW) electric plant at Fort Vrain, Colorado. Commercial power was first generated in 1979. This design used graphite as a moderator and thermal neutrons. The fuel used was a combination of uranium-235 (U-235) with thorium that could produce uranium-233 (U-233). As the first of its kind, the reactor experienced multiple operational problems. These included

---

[1] "High-temperature Gas-cooled Reactor," *Wikipedia* (last modified August 19, 2024). https://en.wikipedia.org/w/index.php?title=High-temperature_gas-cooled_reactor&oldid=1241104713, accessed November 11, 2024.

Nuclear Energy. Edward A. Friedman, Oxford University Press. © Edward A. Friedman (2025). DOI: 10.1093/9780198925811.003.0021

leaking of helium gas, corrosion of metal components, and electrical system issues. While none of these issues posed safety concerns, multiple maintenance problems led to the plant closing in 1989.

Another early HTGR was built at Peach Bottom Township, Pennsylvania from 1966 to 1974. This was a 40-MW electric reactor with a thorium–uranium fuel cycle that produced superheated steam at 995 °F (535 °C). It was judged to be a success in demonstrating the effectiveness of the gas-cooled design but was shut down due to the low power output, which was unable to justify its operational cost.

Project Pele

In recent years, the US Army has initiated a micro-reactor development program, Project Pele, using high-temperature gas cooling with TRISO fuel for a transportable reactor for use in remote locations. This design was motivated by analysis that revealed that the highest casualty rates experienced by combat forces occurred during transport of fuel. The design goal is to have a reactor providing 1–5 MW electric power that would be stable for three years of operation. Given that the Army does not need Nuclear Regulatory Commission approval and has strong Congressional backing, the likelihood of success is quite high, with expectations that this design can be scaled up for commercial applications.

Success in China

In late 2022, China was successful in bringing the world's first modular HTGRs online. These reactors use graphite as a moderator and operate with thermal neutrons. Their plant features two small helium-cooled reactors that use small ceramic-coated fuel elements. The total output of electrical energy is 210 MW. These reactors have the highest output temperatures yet attained and can be used for direct process heat applications. The average output temperature of the helium is 1382 °F (750 °C). It is expected that this design will become widely used in China, as well as actively promoted for the export market. Since with helium gas cooling instead of water cooling, these reactors can be used in locations that are remote from sources of water.

Allegro

An initiative to develop an HTGR is being pursued by a consortium of nations in Central Europe. Participants include the Czech Republic, Hungary, France, Poland, and the Slovak Republic. A multistage program is being pursued in which successive reactor designs are being developed. This project, known as Allegro, is testing various containment materials to explore their potential to withstand high temperatures of up to 1562 °F (850 °C). Various fuels are also being explored.

Japan: Emphasis on Hydrogen Production

The high-temperature engineering test reactor (HTTR) is the first and only HTGR in Japan. The first criticality of the HTTR was achieved on November 10, 1998. The reactor outlet coolant temperature of 1742 °F (950 °C) under full thermal power of 30 MW was achieved on April 19, 2004, for the first time in the world. From January to March 2010, HTTR was operated successfully for 50 days under high temperature and full power conditions.

The Japan Atomic Energy Agency (JAEA), a national research and development agency, and Mitsubishi Heavy Industries (MHI) have been contracted by Japan's Agency for Natural Resources and Energy (part of the Ministry of Economy, Trade and Industry, METI), to conduct a hydrogen production demonstration project utilizing very high temperature, and from 2022 initiated a program to produce hydrogen using an HTTR. Under this program, a newly built hydrogen production plant will be connected to an HTTR owned by JAEA, with the aim of proving the technology for hydrogen production using the high temperature heat obtained from the HTTR.

The determination of specific renovations necessary to connect the hydrogen production plant, along with the licensing procedure, equipment modifications, and testing, will be conducted in stages.

To support future advancements in technologies for hydrogen production, JAEA and MHI will also examine the feasibility of enlarging certain components to allow for large-scale hydrogen production and explore carbon-free hydrogen production technologies in combination with HTGRs.

Industrial Heat Applications

Japan is not alone in pursuing production of hydrogen for high temperature direct heat applications. While hydrogen production is a primary opportunity for HTGRs, there are other significant opportunities for these reactors to reduce carbon emissions. Around a quarter of global energy-related CO_2 emissions originate in direct heating applications that do not involve electricity. These are generally applications for which HTGRs can be engaged for all or part of the processing.

In addition to hydrogen generation, application areas for which HTGRs can play a significant role include: district heating; seawater desalination; pulp and paper production; oil recovery from oil sands and oil shale; oil refining; chemicals production; soda ash production; aluminum production; ammonia production; lime; glass; cement; non-ferrous metal processing; and iron and steel making, among other applications.

Moving forward with applications of HTGRs as a source of industrial heat will require implementing many related designs and establishing new relationships.

In many cases, these direct heat applications must be co-located with the facility with which they are being integrated. Various challenges can arise in that pursuit, including footprints for paired applications that are incompatible with the nuclear plant. Placement in or near residential locations can be another obstacle. Thus, design features must not only be reconciled with existing conditions but also relationships with affected communities must be nurtured. In some cases, relationships involving regulatory agencies will need to be developed.

Another potential bottleneck is the availability of HTGR fuel, which has a higher enrichment level than required for prior generations of reactors. Fuel with an enrichment level of up to 20% U-235, which is what is required, has not been generally available. Up to 2023, the Russian Federation has been the only market source of uranium at this enrichment level. While steps have been taken to expand the availability of highly enriched uranium, success in meeting the levels that might be required for an effective transition to the use of this and other next-generation nuclear reactors could be problematic.

As HTGRs become more widely available, significant expansion of applications in the domain of direct heat industrial applications can be anticipated.

Further Reading

"High-temperature Gas-cooled Reactors and Industrial Heat Applications," *Nuclear Energy Agency*, 2022. https://www.oecd-nea.org/upload/docs/application/pdf/2022-06/7629_htgr.pdf
An in-depth report on high-temperature gas-cooled reactors and their use in industrial heat applications.
"High-temperature Gas-cooled Reactor," *Wikipedia* (last modified August 19, 2024). https://en.wikipedia.org/wiki/High-temperature_gas-cooled_reactor
An overview of the history and status of high-temperature gas-cooled reactor technology.

22

Floating Nuclear Reactors

Interest is growing in installing small modular reactors (SMRs) on floating barges or platforms to provide clean electricity and heat to remote coastal locations, to decarbonize offshore oil and gas or mining activities, or even to provide grid-scale electricity production.

— Lucy Ashton[1]

Reactors for Remote Locations: Overview

The search for portable reactors that could be used in remote locations became an active interest of the US military in the 1950s. Reactors that were developed at that time became precursors of the first US floating reactor designs.

From the first days of nuclear reactor development, the US Army was tasked with developing compact nuclear power plants that could be used in remote locations. That objective was first pursued by the Army Corps of Engineers with the establishment of the Army Reactor Branch at Fort Belvoir, Virginia in 1954. That mission continues into contemporary times with the ongoing development of the Pele advanced reactor.

The first small, easy-to-deploy reactor developed by the Army was a pressurized water reactor (PWR) that generated 10 megawatts (MW) of electricity; activated in 1957, it was designated the SM-1. This reactor was designed by Oak Ridge National Laboratory using standard components that could be mass produced and shipped to remote locations as replacement parts. This development was done in liaison with the Atomic Energy Commission (AEC), which endorsed the initiative, but licensing was not required for this military program. A key characteristic was that it did not require refueling for more than two years. During its 16-year lifetime, the SM-1 was refueled only twice.

The SM-1 became the first reactor connected to an electrical grid, predating the Shippingport Atomic Power Station by nine months. The SM-1 was replicated for use

[1] Lucy Ashton, "Floating Nuclear Power Plants: Benefits and Challenges discussed at IAEA Symposium," International Atomic Energy Agency, November 21, 2023. https://www.iaea.org/newscenter/news/floating-nuclear-power-plants-benefits-and-challenges-discussed-at-iaea-symposium, accessed January 30, 2025.

Nuclear Energy. Edward A. Friedman, Oxford University Press. © Edward A. Friedman (2025). DOI: 10.1093/9780198925811.003.0022

at military bases in Antarctica, Alaska, and Greenland and became the design used in the world's first floating reactor.

The need for a floating nuclear reactor developed at the Panama Canal in 1958 for the control of the water level of locks and a hydroelectric power station at Gatun Lake. The lake is a key part of the Panama Canal and provides millions of liters of water needed to operate the locks each time a ship passes through the canal. When this artificial lake was created in 1913, it was the largest artificial lake in the world. An SM-1 model reactor was enlisted to meet the power shortfall that was required to control the lake's water level.

As part of the SM-1 reactor program at Fort Belvoir, an initiative began in the early 1960s to build a floating unit for remote operations. A retired Second World War cargo ship was modified to provide a hull to house the reactor. Absent a propulsion system, this configuration needed to be towed to whatever location was designated for its use. This system was called MH-1A to designate a mobile high-power reactor. It provided 10 MW electric energy and was named *Sturgis* in honor of an Army general.

In 1967, the *Sturgis* was towed from Fort Belvoir to Lake Gatun, where it provided crucially needed power from 1968 until 1972, facilitating the passage of 15 additional ships through the locks each day.

The successful deployment of the floating nuclear reactor for operation of the Panama Canal stimulated interest in offshore placement of reactors. As most reactors built during the twentieth century required water cooling, their placement was largely adjacent to rivers and large bodies of water. However, suitable locations for reactors were also contiguous to prime residential or recreational locations with residents who objected to reactor development plans. This resistance to placement of new nuclear reactors greatly limited the options of power companies in their planning for new power stations. This was especially true in highly urbanized states.

In 1970, opportunities for development of floating nuclear reactors led Westinghouse to enter a joint venture with Newport News Shipbuilding and Drydock called Offshore Power Systems. This entity established a major construction site at Blount Island, near Jacksonville, Florida, with the objective of building standardized nuclear power plants that could provide energy from the ocean that was economically competitive.

This was an ambitious undertaking with challenging goals. The dock area required was larger than three American football fields. Construction requirements led to the acquisition of the world's largest crane (at a cost of $17 million and a lift capacity of 2 million pounds) to lift the dome of the reactor containment building. More than 1,000 workers were engaged in constructing this facility that was planned to produce four power plants each year.

This developing construction capability led planners at New Jersey Public Service Electric and Gas Company (PSE&G) to consider placing reactors offshore. The undertaking was dubbed the Atlantic Nuclear Power Plant Project. In 1972, PSE&G became a customer of Offshore Power Systems when it placed an order for two 1,150-MW electric reactors. With two identical reactors, operations could be maintained during refueling and other maintenance procedures by shutting down one of the reactors. The plan was to place these reactors on a man-made island anchored in the Atlantic a few miles off the coastline of New Jersey.

To protect this island from severe storms a breakwater was planned that was made of large concrete objects shaped like children's jacks. The plan for the breakwater envisioned about 18,000 "jacks," each weighing 80 tons. These "jacks," known as dolosse, are used in coastal management to dissipate the impact of waves. Dolosse are typically made from unreinforced concrete and are designed to interlock when piled together, dissipating the energy of incoming waves. They are used to absorb the energy of waves and to protect harbor structures and shores from erosion.

The ocean's turbulence required a design in which the breakwater almost surrounded the floating barges. The reactors were planned for placement on large floating barges inside a ring formed by the breakwater. The reactors were designated to be PRWs with electricity transmitted to a shore-based facility via high-voltage submerged cables.

These offshore power plants were considered secure in the event of earthquakes and tsunamis. Given the location of the plant, tsunami waves were not expected to be dangerous.

Offshore Power Systems applied for a license from the NRC to build these reactors in 1973, and the license was finally issued in 1982. During this review period the Three Mile Island accident occurred, which led the NRC to require additional safety features to the design with particular concern with consequences of a severe accident. This concern led to a requirement that material be added that would impede a core melt-through for at least two days.

The proposal for the Atlantic Generating Station stimulated extraordinary concerns for environmental consequences that would result from this man-made island in the ocean and from the cable that would be carrying electricity to a distribution facility on land. Environmental impact analyses were required and organized opposition from environmentalists challenged the installation of an ocean-based power station. PSE&G responded to these concerns by enlisting Ichthyological Associates, a group that originated at Cornell University to study all species of regional marine life and to establish a baseline for the ecological environment. While ichthyology is a branch of zoology relating to the study of fish, this group of scientists examined all species, ranging from the smallest of phytoplankton to the largest of whales, subject to environmental changes that might be introduced by these offshore reactors. They considered the impact of the increased water temperature, the effect of the breakwater, and the cooling system on the environment.

Additional studies were pursued by several teams of experts. These included a team of physical oceanographers who established a baseline on tides, waves, currents, and wind. A group of seismologists and geologists were enlisted to study, among other things, foundation conditions of the ocean floor, geological structures, and regional earthquake history. Deep bores into the ocean bottom showed almost a mile of miscellaneous clays and sands resting on basement rock that had been in place for hundreds of millions of years.

Earthquake concerns were minimal and judged to have consequences only for the breakwater ring since the floating reactors were physically decoupled from the land. The history of tsunamis in the region was found to have caused waves of less than one foot at maximum.

The potential impact of having cables that brought the electric power out of the ocean and overland was studied in detail by general ecologists. *The New Yorker* reported that studies conducted " . . . would eventually consider every tree, every bird, every animal— literally, every mouse—that might in any way be disturbed."

As detailed environmental studies were pursued, plans proceeded for the breakwater ring's construction in 1980 and the floating into place of the first reactor in 1985 and the second in 1987. A key constraint surrounding the selection of site was the need to be within three miles of the coast to avoid intruding in international waters, since the resulting legal issues would be too difficult to resolve. Another factor was the need to support traffic of the vessels that would be towing the floating reactors, where sufficient depth of water was calculated to be a minimum of 40 feet. The location needed to be away from established shipping lanes and contiguous to land that would allow installation of cables. Keeping the water depth close to the 40-foot distance was also significant since the cost of the breakwater could become unmanageable in deeper waters. Nine locations were judged to be suitable with a site 11 miles north of Atlantic City chosen for the first unit of the Atlantic Generating Station.

While development and design activities moved forward, external changes in energy economics and support for nuclear power underwent significant decline. The 1973 oil embargo caused a steep decline in energy demand that continued through the 1970s. A major change in public acceptance of nuclear energy was precipitated by the 1979 Three Mile Island accident. This was followed by President Jimmy Carter placing a moratorium on nuclear power plant construction. These events led to PSE&G canceling their contracts with Offshore Power Systems in 1978 and the dissolution of Offshore Power Systems itself in 1984. The 38-storey crane that was acquired by Offshore Power Systems for $15 million was sold in 1990 to China for $3 million.

Russia's Sea-borne Nuclear Power Plant

Russia has successfully developed small modular reactor (SMR) technology for use in remote locations. A floating nuclear power plant was connected to the grid in December 2019 in an isolated town across the Bering Strait from Alaska. Named the Akademik Lomonosov, it is offshore at the town of Vilyuchinsk in far-eastern Russia. The energy is generated from two pressurized light-water reactors with each producing 35 MW electrical. These are like reactors used in Russian icebreakers. This power plant is the world's most northern reactor location, and its units can operate for from 3 to 5 years without refueling.

Rosatom, the Russian company that produced these reactors, sees the Akademik Lomonosov as the first of a series of SMRs that can be located offshore at various locations worldwide. Rosatom is planning to build units with a range of power capacities to meet what they see as a large potential export market. They plan to develop a mass market for this class of reactors. Five floating plants are planned for the Russian Arctic region. Rosatom is in discussion with 15 countries for sale of floating reactors, including Malaysia, Indonesia, and Argentina.

Aborted Chinese Plans for Floating Nuclear Reactors

China engaged in ten years of floating nuclear reactor development and then abruptly terminated this initiative in 2023. The reason for this was most likely political as China was planning to deploy floating reactors in the South China Sea, which is a potential conflict zone. In the event of hostilities, a floating reactor could be a target for drones. Other scenarios involving attacks on a floating reactor are also possible.

Seaborg Technologies

Seaborg Technologies is a private Danish company that is developing molten-salt reactors for deployment on barges. Unlike other similar reactors, the Seaborg design does not use graphite as the moderator. Instead, it uses sodium hydroxide contained in pipes adjacent to and interlaced with those containing the molten fuel salt. This configuration allows a more compact design, hence the reactor's name, Compact Molten Salt Reactor.

The safety system employs a frozen salt plug, which melts in case of overheating and allows the reactor's contents to flow into an adjacent container where a chain reaction cannot be sustained. Seaborg plans to place the reactors in containers for placement on barges. Single reactor output is planned to be 100 MW electric with the possibility of a barge accommodating up to eight more reactors.

Seaborg is part of a consortium that includes South Korean shipbuilder Samsung with plans for placing barges along the coastlines of Norway and Indonesia. Seaborg envisions having operational units in 2030. In addition to applications for electricity production, direct heat applications for desalination, hydrogen production, and metals processing are also planned.

In addition to Rosatom and Seaborg, other nuclear reactor companies will likely compete for a promising export business for floating reactors. Two entries into this field are NuScale and ThorCon. NuScale plans to build small modular nuclear reactors using combinations of unique tubes containing pressurized water reactors that each produce 77 MW electric. They have partnered with Prodigy Clean Energy of Canada for construction of floating reactors.

ThorCon plans to float barges with two 500-MW electric reactors that use a molten-salt design with thorium as the basic fuel. These are scaled-up versions of reactors that were first developed at Oak Ridge National Laboratory in the 1960s. ThorCon envisions shipyard construction of these systems with an ambitious goal of completing 20 units per year. ThorCon is partnered with the Indonesian government, which plans to take advantage of its large deposits of thorium.

Seaborg, NuScale, and ThorCon are also discussed in Chapter 23, which covers SMRs.

Further Reading

"Floating Nuclear Power Plant," *Wikipedia* (last modified September 26, 2024). https://en.
   wikipedia.org/wiki/Floating_nuclear_power_plant
A summary of floating nuclear reactor developments.
Lucy Ashton, "Floating Nuclear Power Plants: Benefits and Challenges Discussed at IAEA Symposium." International Atomic Energy Agency, 2023. https://www.iaea.org/newscenter/news/
   floating-nuclear-power-plants-benefits-and-challenges-discussed-at-iaea-symposium
*A short summary of the symposium on floating nuclear reactors that was held on 14–15 November 2023
   by the International Atomic Energy Agency.*

23

Small Modular Reactors (SMRs)

Progress since the publication of the inaugural volume of the NEA Small Modular Reactor Dashboard has been rapid and is accelerating, with multiple projects moving from conceptual design, licensing and siting to breaking ground on construction.

— Nuclear Energy Agency Director-General William D. Magwood, IV[1]

Small Modular Reactor (SMR): Overview

The overwhelming evidence of global warming and dramatic changes in weather patterns have forced new thinking about energy. Also, conflicts around the world have led to uncertainties regarding reliable access to fossil fuel supplies and forced reexamination of energy policies and practices. In the search for stable and affordable energy, there is now greater interest in small modular nuclear reactors (SMRs).

While it is useful to categorize nuclear reactor development as Generation III and Generation IV, examining public policy and tracking technological development through an SMR lens helps place issues into perspective.

A driving force for the development of SMRs is the prospect of achieving designs that allow production-line construction of replicable units. The nuclear industry has been plagued by one-of-a-kind construction that has been both expensive and invariably associated with overly long production schedules. The prospect of lowering costs and eliminating delays has attracted extraordinary interest from industry and governments around the world. In 2023 there were more than 80 SMR designs under development in 19 countries.

The concept of factory construction is necessarily linked to size limitations since transportation from the factory to an installation site needs to be managed in a safe and economical fashion. Transportation of reactor components by truck, rail, and ship must be incorporated into design plans.

---

[1] William D. Magwood IV, "G7 Emphasises the Role of Nuclear in the Energy Transition," Nuclear Energy Agency (NEA), May 8, 2024. https://www.oecd-nea.org/jcms/pl_92773/g7-emphasises-the-role-of-nuclear-in-the-energy-transition, accessed January 30, 2025.

Nuclear Energy. Edward A. Friedman, Oxford University Press. © Edward A. Friedman (2025). DOI: 10.1093/9780198925811.003.0023

In addition to the prospect of more manageable costs and production timescales, the inherent safety of Generation-III and Generation-IV designs for SMRs has captured the attention of planners.

The promise of fail-safe operations has made new nuclear technologies extremely attractive. For example, if there is overheating due to an accident, a fail-safe system will automatically initiate cooling processes. This cooling takes place not because of any human intervention and not because of any technological intervention, but through the forces of nature. An attractive design feature is the automatic circulation of cooling water, which saves the day.

NuScale

NuScale is an innovative small modular company that is making significant progress in the field and provides a good example of what's happening in this world of SMRs. While NuScale may or may not succeed in implementing functioning reactors, the strategy they are pursuing provides a clear picture of design features needed for success in this domain. A basic requirement is based on the need to transport functional reactor units from the factory where they are built to the site where the reactor will operate. Since the only way to transport them is by rail, truck, or ship is in containers, the size of a shipping container therefore becomes a constraint that must be recognized. Since typically the largest container that you can put on a truck is about 75 feet long by about 10 feet wide (23 meters long by about 3 meters wide), those dimensions dictate the size of the reactor.

Next, we need to review the components of the working element of the nuclear reactor. At the center there is a reactor core, which generates heat by sustaining a chain reaction. That heat is carried by a fluid to a secondary circulating fluid that produces steam, which operates a turbine that generates electricity. The reactor and turbines that produce electricity are placed in a containment enclosure that protects public areas.

The NuScale reactor is in a cylinder about 75 feet high and about ten feet in diameter (Figure 23.1). The cylinder contains all the components for a nuclear reactor. This design is that of a pressurized-water reactor (PWR), but it is missing the cooling pumps that are generally included in the make-up of a PWR. Since most reactors in operation today are PWRs, it is a well-known technology. Note that the reactor core, which produces the heat, is at the very bottom of this cylinder.

If an accident causes the reactor to heat over its planned limits, automatically, through the forces of gravity, water will start circulating. Cold water travels down in the reactor and hot water rises. This fail-safe behavior is activated by the laws of physics. It does not require pumps, devices, or actions by operators!

The NuScale design anticipates that an individual 75-foot (23-m) cylinder will produce 77 megawatts (MW) electric. To provide the energy of a large power plant that typically produces 1000 MW, 12 cylinders need to be configured as an integrated power plant. The total output of such a plant, which would provide 924 MW electric, is the stated goal of the NuScale design.

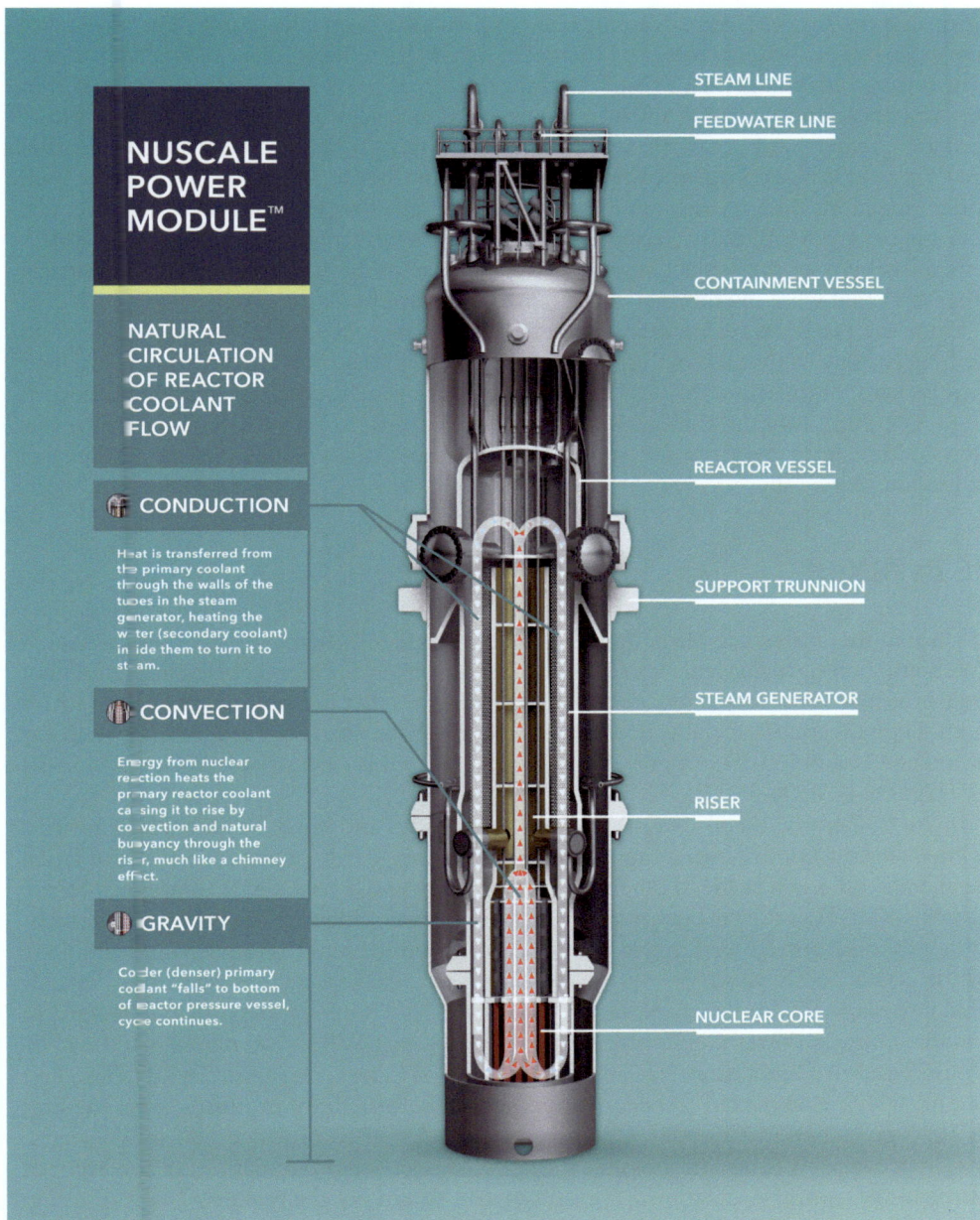

Figure 23.1 *Proposed NuScale reactor design schematic.*

All this fits in 14 hectares (about 0.14 km), which is an area equivalent or smaller than existing fossil-fuel plants, and the ability of SMRs to fit on this existing footprint is an enormous advantage. This ability to transition from fossil fuel to green energy within the same location is a major advantage over wind and solar possibilities, both of which need far greater land areas. Proximity to the existing infrastructure also obviates the need for grid connections to bring energy from another location. In many cases such grid connections do not exist and if they do they entail energy losses and extra expenses.

A major opportunity for streamlined transitions to green energy provided by SMRs is the possibility of the replacement SMR plant's use of the same turbines and electrical generators that were operational in the fossil-fuel plant being reused.

Since the NuScale design incorporates PWR technology, it was able to capitalize on the long history of that technology in securing licensing approval from the US Nuclear Regulatory Commission (NRC). Approval of their Final Safety Evaluation Report in August of 2022 was the first such approval by the NRC of an SMR. That approvals allows NuScale to plan reactor construction in the United States and to market this design internationally.

GE BWRX-300

Given that there is an established track record for safe, effective, and economical use of PWRs and boiling water reactors (BWRs), the most likely initial success in implementing small modular reactors will most likely be with reactors that use water as the heat transfer agent. High on the list of viable SMRs are those that have been pursued for many years as advanced Generation-III designs. A leading reactor in this category is the Generation-III GE–Hitachi BWRX-300 BWR.

In 2007, General Electric (GE) of the United States and Japan's Hitachi Ltd. formed a global alliance combining elements of their nuclear power businesses. Outside of Japan the alliance is known as GE–Hitachi Nuclear Energy and is based in Wilmington, North Carolina. GE–Hitachi has been actively pursuing development of the boiling water technology that GE introduced in the 1950s as candidate for mainstream SMR technology for the 2030s and beyond.

Major plans for implementation of BWRX-300 technology are moving forward in Canada. Ontario Power Generation plans to build four BWRX-300 reactors at the Darlington Nuclear Generating Station on the shores of Lake Ontario in Canada. The first reactor, which is expected to become operational before 2030, promises to be the first SMR to go online in Canada. This reactor is designed to produce 300-MW electric using a proven BWR design that incorporates fail-safe water circulation. In the event of overheating, natural circulation will result in safe cooling of the reactor.

Plans for implementation of this reactor in Europe are also developing with the largest contract anticipating construction of ten reactors in the early 2030s in Poland. This project, pursued with the Polish firm Synthos Green Energy (SGE), would be a breakthrough for nuclear collaboration between firms in North America and Central Europe.

BWRX is a mature technology that has its origins in 1950's reactor development by the Argonne National Laboratory and GE. Their joint innovative work resulted in a demonstration in Vallecitos, California of a 5-MW electric reactor, which was the first privately developed nuclear reactor to deliver energy to a public grid. The success of the Vallecitos reactor led to the development of the 180-MW electric BWR in 1956, the Dresden 1 in Morris, Illinois. BWRs have been subject to continuous improvement by GE since the 1950s. A significant milestone was achieved in 2010 when the NRC issued design certification for the Economic Simplified Boiling Water Reactor (ESBWR) that employed passive safety systems and a simplified design using natural circulation. These attributes allow the reactor to cool itself for more than seven days without operator intervention or the need for on- or off-site power.

This Generation-III+ design is based on the natural flow in the core of hot water, which is less dense than cold water. Water therefore circulates with the hot water rising and the cold water flowing to the bottom of the reactor vessel without the need for circulation pumps. The construction of reactors that take advantage of this natural phenomena have intrinsic simplicity that facilitates less expensive construction that can be accomplished more quickly than earlier designs.

The BWRX-300 design evolved from that of the ESBWR and while it presents itself as an innovative technology it is based on many years of experience and careful planning. The road toward construction approvals in many countries should therefore be facilitated by having a record of proven technology. There have been over 100 BWRs built and operated around the world. More than half of these were built by GE. In the United States roughly a third of operating reactors are BWRs, with all operational reactors having been constructed by GE.

Westinghouse AP300

Westinghouse did not enter the SMR market until May 2023 when it announced its intention of producing a scaled-down version of its 1000-MW electric AP1000. The planned AP300 that would produce 300 MW electric is on a timeline for NRC licensing approval by 2027 and construction beginning in the early 2030s. Given the proven experience with the AP1000, this schedule is likely to be implemented in a timely fashion.

ARC-100

The ARC-100 is a Generation-IV 100-MW electric sodium-cooled fast reactor (SFR). It leverages proven SFR experience, particularly the Experimental Breeder Reactor (EBR-II) reactor that was operated successfully for 30 years, from 1964 until its decommissioning in 1994. Operations and detailed studies were conducted on the EBR-II by Argonne National Laboratory, including a dramatic demonstration of its safe design that was conducted before an audience of international observers on April 3, 1986. That event witnessed the cooling system cutoff while disabling the system's control rod safety response. The reactor cooled down safely due to its natural circulation of its components. Other safety demonstrations that removed the system components to which heat circulated were also conducted. The EBR-II's development was terminated due to political factors, despite its promising performance achievements.

Canada's ARC Clean Technology was formed to resurrect the earlier work of Argonne National Laboratory in pursuit of this Generation-IV SMR design. In 2018 they entered a partnership with New Brunswick Power to build an ARC-100 reactor at the Point Lepreau Nuclear Generating Station in New Brunswick, Canada, with a target date for completion of 2029. A goal for the updated version of the EBR-II is the ability of the current design to burn spent nuclear fuel.

In 2024, Korea Hydro and Nuclear Power (KHNP) joined ARC Clean Technology and New Brunswick Power to form a three-way partnership, the objective of which is to engage in building and marketing large numbers of the ARC-100 globally. KHNP is the largest electric power company in Korea and one of the world's largest nuclear operators. The company has acquired world-class capabilities in the construction and operation of nuclear power plants globally.

The "walk away" passive safety system that ensures that the reactor will not melt down, even in a disaster that causes a complete loss of power, makes ARC-100 extremely attractive to prospective users. In addition, it can be fueled with the nuclear waste produced by traditional reactors and its 20-year refueling cycle offers significant levels of proliferation resistance.

French Nuward Initiative

In 2023, *World Nuclear News* reported that Electricité de France (EDF) is embarking on a large-scale project to capture the SMR market in Europe and globally through development of a reactor known as Nuward. This is a high-profile initiative embedded in national planning. EDF strongly believes that SMRs have the potential to play a crucial role as part of the global initiative to curb the impact of climate change. The Nuward SMR offers a sustainable solution for quick access to baseload, dispatchable, affordable low-carbon power generation. Plans call for a 340-MW electric power plant consisting of two independent 170-MW electric units that incorporate Generation-III+ PWR technology. The design anticipates modular units that are mass produced in factories. They would incorporate flexible power output capabilities that would allow their integration into networks having wind and solar generators that have variable output. Non-electric and direct power applications are also anticipated. A key strategic objective is their use in replacing aging coal-fired plants in the 300–400-MW range. A 24-month operating cycle is planned with a design life of 60 years. EDF is pursuing a timetable where 2030 would witness the first operational realization of this reactor.

UK Rolls-Royce Program

Rolls-Royce in the UK plans call for the construction of a 470-MW electric reactor that is much larger than the size of nominal SMRs, although it shares the SMR design feature of being factory built and boasts a footprint that is one-tenth of the size of a conventional nuclear generating site. The Rolls-Royce SMR incorporates proven Generation-III+ PWR technology. Rolls-Royce is the only nuclear reactor developer in the UK and boasts an experienced workforce that has provided nuclear reactors for the British Nuclear Navy since 1993.

Small Modular Reactors—Power Output Below 100 Megawatts

Many initiatives for the development of micro-reactors are being pursued throughout the world. This category of reactors with energy output below 100 MW are sought after for use in remote communities, mining centers, disaster locations, and isolated industrial facilities. The interest in this category of power source is high with a wide range of operating temperatures depending on the intended application.

The highest temperature reactor in this category is the High Temperature Engineering Test Reactor (HTTR) that achieved an outlet temperature of 1742 °F (950 °C) in 1998 in Japan. This reactor provided 30 MW of output thermal power for use in hydrogen production and other direct heat applications such as desalination. Other countries are developing similar reactors and seeking collaboration with Japan for export contracts.

The American and Canadian nuclear energy company BWXT is pursuing development of a high-temperature gas-cooled (HTGR) 50-MW thermal output reactor known as the BWXT Advanced Nuclear Reactor (BANR) that would be a modular system which would be factory-fabricated, small, and light enough to be transported via rail, ship, or truck. Applications include industrial process heat and electrical production at remote locations, including military sites that are at high risk to attack when receiving fuel shipments.

A water-cooled micro-reactor that takes advantage of the more than 60 years of experience with this technology by Westinghouse is the eVinci. This 5-MW electric reactor is under development. It is designed to run for eight or more years at full power without refueling and to be fully factory assembled. It is designed for shipment via rail, barges, and trucks. Its ability to load-follow and load-shed within milliseconds makes it suitable to help maintain a constant energy output when paired with a wind or solar base plant. Heat transfer in an eVinci reactor is processed by a heat pipe. This innovative mechanism uses a volatile liquid in contact with a solid surface that conducts heat and turns the liquid into a vapor that travels along the heat pipe to a cold interface, where it deposits heat and transforms back into a liquid. The liquid then returns to the hot interface through gravity and the cycle repeats. Incorporating heat pipe technology greatly simplifies design and eliminates numerous components that are found in traditional

PWRs. Of great importance is that risk from maintaining water under high pressure is eliminated, as is the risk of a loss of coolant accident.

Development is taking the place of a 35-MW thermal demonstration reactor known as Hermes, by Kairos Power at Oak Ridge, Tennessee. This is part of a plan to construct a 140-MW electric commercial reactor that uses fluoride salt as a coolant. This fluoride salt-cooled reactor will be the first non-water reactor to be built in the United States in more than 50 years. In December 2020, Kairos received a Department of Energy (DOE) Advanced Reactor Demonstration Program award for its participation in this innovative program. The total cost was $629 million over seven years with the DOE contributing $303 million. The fluoride salt coolant is extremely stable at elevated temperatures and has excellent heat transfer characteristics. The Hermes reactor's stability is further enhanced using TRISO fuel pellets that retain the nuclear waste within the ceramic capsule. Construction is proceeding in collaboration with the Tennessee Valley Authority. With an output temperature of 1202 °F (650 °C), the Hermes reactor can be used to provide direct transfer heat for chemical production processing and ammonia refining.

A joint venture known as Global First Power Ltd. (formed by Ultra Safe Nuclear Corporation based in Seattle, Washington and Canada's Ontario Power Generation) is pursuing development of the Micro Modular Reactor (MMR) in Chalk River, Ontario, Canada. This HTGR uses TRISO ceramic fuel to produce 45-MW thermal heat that could have multiple applications over a wide range of temperatures up to 1652 °F (900 °C). A molten-salt reservoir is used to store output heat and transfer it as needed for use in applications. The modular units are designed for factory construction with site installation that is suitable for scalability that incorporates multiple units. MMR technology is being licensed in the United States and Canada with a marketing goal of worldwide implementation. Ultra Safe is pursuing implementation plans for MMR technology with the Manila Electric Company in the Philippines.

The Research Centre Řež in the Czech Republic is developing an SMR known as the Energy Well. This 20-MW Generation-IV reactor uses fluoride salts as a coolant with TRISO fuel and is designed for transport in a container that can be easily shipped via road or rail. The power plant is configured with three physically separated units. The primary heat production unit is physically distinct from an intermediary heat transfer unit. The reactor core is planned for placement in a pit below ground for safety and security. The heat generation unit is designed to operate for seven years and then be moved into a holding area while a fresh heat generation unit is installed to ensure continuous operation of the plant. After being stored for two years, it is removed. A demonstration of Energy Well is planned for 2030.

An aggressive approach to nuclear energy development is being taken by Last Energy, located in Washington, DC. Last Energy is entering into contracts for small reactors that use standard PWR designs. The reactor, known as PWR-20, is a PWR operating at a modest 572 °F (300 °C) output temperature with a power output of 20 MW electric. It is slated to occupy an area of less than an American football field. Last Energy is entering into power purchase agreements (PPAs), which provide for the provision of electrical energy for a given period—between 5 and 20 years. Last Energy, a founding member of the Texas Nuclear Alliance, is contracting with a company in Haskell County, Texas to fabricate the SMR using proven designs with off-the-shelf available components. The reactor is housed in nine modular units, each about the size of a shipping container, allowing for transport to the site where they would be used where assembly into an operating unit is implemented. Safety is ensured by below-ground burial of this unit connected to control and energy distribution mechanisms in a connecting above-ground facility. As of early 2024, contracts have been signed for installation of 34 units with ten units in Poland and 24 units in the United Kingdom. Initiation of this development work is contingent upon approval from the cognizant European oversight authorities.

Illustrative of the diversity of small modular nuclear reactor development taking place throughout the world is the Central Argentina de Elementos Modulares (CAREM) reactor in Argentina. This 32-MW electric reactor is Argentina's first domestically designed and developed nuclear power reactor. CAREM is a PWR design using natural circulation that provides inherent safety against core meltdown. Construction of this unit was first proposed in 1984, but postponement resulted in a timetable for completion around 2027.

Silicon Valley startup Okla is pursuing development of a fast reactor that does not use a moderator to engage the heavy elements present in nuclear waste in a heat-producing nuclear chain reaction. Known as Aurora, the micro-reactor under development would produce 1.5 MW electric. However, Okla's application to build this plant at Idaho National Laboratory was denied by the NRC in 2022 because of gaps in the description of its safety systems and components. Given that this liquid sodium-cooled reactor design was operational as an Argonne National Laboratory development project from 1964 to 1984 leads to expectations that resubmission of an application to the NRC for operations for development to proceed is likely to be forthcoming.

Jimmy Energy is a French company developing a Generation-IV HTGR using helium gas as the coolant. Implementation is hoped for in 2026 with heat output of the Jimmy reactor to be 1112–1382 °F (600–750 °C). These reactors are not intended for production of electricity but rather for industrial heat applications. The design calls for the use of TRISO fuel that is enriched to just under 20% uranium-235. In the event of overheating the reactor would shut down and the graphite moderator would passively dissipate heat.

The smallest operational nuclear reactors have been in use since 1974 at the Bilibino Nuclear Power Plant in Chukotka, Siberia, Russia, where they have provided energy for gold mines. They are slated for replacement but continue to operate into the mid-2020s with a power output of 12 MW electric. Three such units provide heat, desalination, and electricity for this remote location.

Swedish company LeadCold Reactors AB is developing a lead-cooled Generation-IV SMR for mass production in the 2030s. It is building a demonstration reactor that has 55-MW electric capacity that could operate for 10–30 years without refueling. Also, the Swedish Advanced Lead Reactor Demonstration (SEALER-D) is being developed in collaboration with the municipality of Studsvik. This lead-cooled design takes advantage of having a coolant fluid that does not interact violently with other liquids and materials, as does liquid sodium. Its high boiling temperature of 3164 °F (1740 °C) eliminates loss of coolant through boiling accidents. SEALER uses 12% enriched uranium.

Summary

This wide assortment of SMRs presented here illustrates the intense interest in this technology that is found in countries around the world. With many diversified designs aiming to achieve implementation of a demonstration model by 2030, the chances of seeing such a development come to fruition are high.

Further Reading

"NEA Small Modular Reactor Dashboard," 2nd edn. *Nuclear Energy Agency*, 2024. https://inis.iaea.org/records/34gey-2bs04

A comprehensive compilation of SMR developments taking place in countries throughout the world.

"Small Nuclear Power Reactors," *World Nuclear Association* (last modified February 16, 2024). https://www.world-nuclear.org/information-library/nuclear-fuel-cycle/nuclear-power-reactors/small-nuclear-power-reactors.aspx

In-depth report on status of SMR developments worldwide.

24

Nuclear Reactor Export Market

By many measures, Russia's state-controlled nuclear energy company, Rosatom, has primacy in the global nuclear energy market.

— Marina Lorenzini and Francesca Giovannini[1]

Export Market: Overview

The export of nuclear reactors is a multi-faceted geopolitical transaction. At the top of the list of connections is the economic arrangement that can provide leverage in long-term relationships if significant credit is extended in the purchase. Auxiliary factors include the ongoing supply of nuclear fuel, the management of nuclear waste, the training of nuclear personnel, and the management of plant operations. Russia and China have been aggressive in using nuclear exports to gain ongoing influence through these multiple relationships.

As of 2024, the following 12 countries have constructed nuclear reactors: United States, Russia, France, China, United Kingdom, Canada, India, Japan, South Korea, Germany, Sweden, and Argentina. Of the 440 reactors that are either operational or could be functional but whose operations have been suspended, 335 are in these 12 countries. Of the operational reactors roughly a third have been acquired in the export market (Bowen, 2023).

In 2024, there were 58 nuclear reactors being constructed with 43 projects undertaken in the 12 reactor production countries. By far the most active development site is China, with 23 reactors under construction, followed by India with eight.

---

[1] Marina Lorenzini and Francesca Giovannini, "Five Reasons that Russia's Nuclear Exports Will Continue, Despite Sanctions and the Ukraine Invasion. But for How Long?" *Bulletin of the Atomic Scientists*, May 17, 2022. https://thebulletin.org/2022/05/five-reasons-that-russias-nuclear-exports-will-continue-despite-sanctions-and-the-ukraine-invasion-but-for-how-long/, accessed January 31, 2025.

Nuclear Energy. Edward A. Friedman, Oxford University Press. © Edward A. Friedman (2025). DOI: 10.1093/9780198925811.003.0024

Russian Exports

By far, Russia is the most active country engaged in the export of nuclear reactors. Of the 15 projects in 2024 that can be categorized as export programs, 11 are Russian (Schepers, 2019). They are building four reactors in Türkiye, four in Egypt, two in Bangladesh, and one in Iran.

In 2016 Russian-owned Rosatom concluded a contract with the government of Bangladesh for construction in the village of Rooppur of two VVER-1200 nuclear reactors that will provide 2.4 gigawatts (GW) electric, which is 15% of the country's total. This was the culmination of discussions with various potential sources since 1961. The final agreement called for Rosatom to provide 90% of the financing with the facility going online in 2024. This $13 billion project engaged 10,000 employees from Bangladesh and 2,500 specialists from Russia. The vast scope of this undertaking is testimony to the capacity of the Russian Federation to successfully engage in such projects simultaneously in several countries across several continents.

In April 2023, Russian President Vladimir Putin personally participated in a ceremony that was conducted remotely to mark the arrival of the first fuel at the under-construction Akkuyu nuclear power plant in Türkiye. This is the first of three plants that are under contract with Türkiye. This 1200-megawatt (MW) VVER plant provides 11% of the electricity used in Türkiye. This reactor program is evidence of the deep ties that exist between the two countries, which presents a relationship that is of concern to the West.

A major Russian initiative is underway in Egypt where four VVER pressurized water reactors (PWRs) are under construction at El Dabaa on the Mediterranean coast 250 km west of Alexandria. This complex is planned to provide 4.8 GW electric that would satisfy 50% of Egypt's electrical power requirement. These plants have anticipated commissioning dates that start in 2026 with completion scheduled for 2030. Fulfillment of this program would see realization of plans that were first announced in 2007 but experienced delays due to political turmoil and local opposition.

Eighty-five percent of the financing of the El Daba program is provided by a long-term low-interest loan from Rosatom with the remaining funds being raised locally. The plant itself would be owned by the Nuclear Plant Authority of the Arab Republic of Egypt. Rosatom is contracted to oversee operations of the plant and management of its nuclear fuel.

Russia is engaged in an aggressive marketing campaign to sell nuclear reactors to countries in Africa. Russia seeks to tie African countries to financing plans in which it lends the target country a considerable portion of the purchase price of the reactor. Russia also offers contracts for fuel supply and management with provisions to bring spent fuel back to Russia for processing.

Russia has had a checkered relationship with South Africa, the continent's most industrialized country, with respect to nuclear energy. In 2014 it signed a deal worth an estimated $76 billion: the deal was shrouded in secrecy and was later canceled by a court judgment for being illegal and unconstitutional. It had been negotiated by the government of Jacob Zuma, who was later indicted for corruption. In 2023 a deal was struck in which a Russian bank would finance the restart of a gas-to-liquid oil refinery that had been out of operation for several years. This is part of a larger energy negotiation that could lead to acquisition of a second South African nuclear power plant from Russia. The geopolitical context for these negotiations is being shaped by the sanctions imposed on Russia by the United States due to the war in the Ukraine. The Russian nuclear firm Rosatom is emphasizing promotions in countries like South Africa that have adopted a neutral position with respect to Ukraine.

Several other African countries have been in negotiations for nuclear technology with Russia. Included is Nigeria, Africa's largest country by population and biggest in terms of GDP. Discussions have also been pursued with Ghana, which is seeking a nuclear power plant. Rosatom has signed cooperation agreements with both countries and is also pursuing negotiations with Ethiopia and Zambia.

In 2023 Russia also signed a memorandum of understanding for development of a nuclear reactor for the country of Burkina Faso, a country where only about a quarter of the population has access to electricity.

These negotiations and agreements are taking place within a strategic context in which Russia is creating a broad conceptual framework for nuclear power policy. In July 2023, Rosatom organized the Second Summit and Russia–Africa Economic and Humanitarian Forum which hosted the plenary session "Nuclear Technologies for Development in Africa."

As part of Russia's comprehensive program to promote nuclear technology exports it maintains training opportunities for support staff from the target export countries. Rosatom reports that since 2018 more than 2,800 specialists from China, Türkiye, Egypt, Hungary, Belarus, and Bangladesh have completed training programs covering a wide variety of roles at nuclear power plants. These sessions include presentations on how to manage VVER-1200 power units safely and effectively. Enrollments in the 2023–2024 period were in the range of 500 foreign specialists annually.

In Europe, Russia continues to expand its role as a provider of nuclear energy with mixed success. Bulgaria, which operates nuclear reactors from Russia acquired during the era of Soviet domination, is turning toward Western providers, while Hungary, which has four Soviet-era reactors, is moving ahead to acquire two additional units. EU sanctions against Russia that were imposed due to the war in Ukraine do not apply to nuclear energy.

Chinese Exports

China's nuclear export program is an integral component of its massive Belt and Road Initiative (BRI), sometimes referred to as the New Silk Road. This ambitious infrastructure project was initiated in 2013 by President Xi Jinping as a program to extend China's influence in multiple locations throughout the world, including countries in Africa, Oceania, and Latin America. Developments envisioned for this New Silk Road include a vast network of railways, energy programs, highways, and telecommunications structures. Sites designated for engagement are located southward to Pakistan and into Southeast Asia. Plans also call for a twenty-first-century maritime Silk Road stretching from the Indian Ocean to Africa and Europe. A total of 147 countries have entered into agreements with China or have indicated interest. Through loans and development programs, Beijing seeks geopolitical leverage through the BRI. For example, after joining BRI in 2022, Nicaragua severed ties with Taiwan in favor of relations with China.

An integral component of BRI is the export of nuclear reactors with approximately 30 nuclear reactors slated for export to BRI countries by the mid-2030s. As of mid-2020s, the most extensive implementation of Chinese nuclear export activity has been in Pakistan where four 300-MW electric reactors have been installed and two 200-MW electric nuclear reactors are contracted for Karachi. Groundbreaking for the first unit took place in September 2023.

China has been in nuclear energy collaboration discussions with numerous countries as they pursue BRI implementation, including South Africa, Kenya, Egypt, Sudan, Armenia, and Kazakhstan.

An agreement that underwent change due to conflicting East–West objectives took place in 2018 when Romania entered into an agreement with China for the construction of 1200-MW electric reactors at the Cernavodă Nuclear Power Plant. These would be in addition to two CANDU 700-MW electric plants that went into operation in 1996 and 2007 through contracts with Canada. However, the arrangement with China was clouded when the US Justice Department accused the China General Nuclear Power Corporation of nuclear espionage. The US justified this accusation, citing "conspiracy to unlawfully engage and participate in the production and development of special nuclear material outside the United States, without the required authorization from the US Department of Energy," https://www.justice.gov/archives/opa/pr/us-nuclear-engineer-china-general-nuclear-power-company-and-energy-technology-international. This was, in effect, an allegation of theft of US nuclear technology for military purposes. The United States blacklisted the China General Nuclear Power Corporation, which led to Romania canceling its agreement with China in June 2020. It then ratified an intergovernmental agreement with the United States for completion of the work started by China on the two new units at Cernavodă.

South Korean Exports

South Korea has become a major competitor in the nuclear export market. In an impressive display of competence, South Korea contracted for construction of four nuclear power plants in the United Arab Emirates (UAE).

In December 2009, the Emirates Power Corporation contracted with a consortium led by the Korea Electric Power Corporation (KEPCO) to install four 1400-MW electric power plants. In 2012 KEPCO began construction of Barakah Unit 1 at the Barakah Nuclear Energy Plant, which became operational in August 2020. This on–budget and on-time performance was implemented with units 2, 3, and 4. Barakah Unit 4 entered service in 2023.

Following the successful reactor development in the UAE, KEPCO signed contracts with the government of Poland for construction in the Pątnów, Konin region of Poland of two APR1400 reactors, with the same design features as implemented in the UAE. Launch of the first of these power plants in 2035 is anticipated. This action triggered legal action on the part of Westinghouse, which claimed that their intellectual property agreements with South Korea prevented South Korea from engaging in such exports without approval from Westinghouse. This complex legal action will not be resolved quickly, but it has not caused any delay as of mid-2024 in the work being undertaken in Poland. Further, KEPCO is actively engaged in pursuing export contracts with the goal of exporting 10 nuclear plants by 2030.

French Exports

France is not only a leading consumer of nuclear energy, but it also actively engages in exporting reactors. It has a record of exporting reactors to China, Finland and the United Kingdom. Early exports occurred before 2000 with installation of 1000-MW Generation-II+ reactors in China that were connected to the grid in 1994 and 1995. Subsequently France has been exporting a 1750-MW Generation-III design known as the European Pressurized Reactor (EPR). Two EPR reactors became operational in China in 2018 and 2019. China used experience gained with French technology to initiate domestic development of nuclear technology.

Construction of an EPR reactor began in 2005 in Finland but it experienced successive problems and delays with regular electricity production only in 2023. Problems experienced with the EPR implementation in Finland prompted France to redesign the EPR reactor. A simplified design known as EPR-2 has evolved from this process.

Two new EPRs are under construction at Hinkley Point C nuclear power station in Somerset, England. This project, which began with joint funding from China, needed adaptation in 2023 when China withdrew from the program due to international relations tensions. The French company Electricité de France remains as the sole funder. A start date for these reactors of 2031 is anticipated.

The French government announced in 2021 that it would be promoting the development of Small Modular Reactors (SMRs) primarily for export that would become available around 2030. A strong market is anticipated for SMRs for the period from 2030 through 2050.

US Exports

The first exports of US nuclear reactors were General Electric (GE) Mark 1 models that were built at the Fukushima Daiichi Nuclear Power Plant. Five of the six reactors at Fukushima were these GE boiling water reactors (BWRs) with the first three coming online between 1971 and 1975. These reactors provided power of 460 MW electric, 784 MW electric, and 1100 MW electric. The design of the containment enclosure for these reactors was questioned by three GE engineers who quit their jobs with GE in protest over this design choice. The containment shell of these BWRs was cheaper and easier to build because of its thinner and smaller containment shell over the reactor vessel. The events at Fukushima witnessed the rupture of three of these containment shells due to hydrogen explosions. As discussed in Chapter 13, that accident was due to inadequately protecting the backup generators from the tsunami whose height exceeded expectations.

GE also provided two BWRs to Taiwan that entered operations in the 1980s and were decommissioned after 40 years of service. Taiwan also acquired units for two power plants from Westinghouse in the mid-1980s. These PWRs are also scheduled for decommissioning in the mid-2020s. Following the Chernobyl and Fukushima accidents Taiwan experienced sustained anti-nuclear protest movements that have influenced public policy.

Westinghouse constructed two PWRs in Brazil. The first was a 609-MW electric unit that became operational in 1985 and the second a 1275-MW electric unit that became operational in 2000. Plans for a third unit have been delayed due to a shortfall in investment capital.

GE installed two 775-MW electric reactors in Mexico. The first became operational in 1989 and the second in 1994. Negotiations for these reactors and construction took place in the 1970s prior to the Chernobyl accident.

Westinghouse constructed a 688-MW electric PWR plant in Slovenia that is jointly owned by Croatia. This was the first Western reactor to be introduced into Eastern Europe. Construction started in 1975 and connection to the grid took place in 1981.

The Philippines acquired a 621-MW electric PWR from Westinghouse that was completed in 1984 but was never operational. Due to safety concerns related to earthquakes fuel was never loaded into the plant. The plant's status was on hold at the time of the Chernobyl accident, which led to the unit being mothballed. The plant continues in this status with the government of the Philippines absorbing an expense of nearly $1 million a year for maintenance.

These citations of US exports by GE and Westinghouse were all initiated in the 1960s and 1970s with construction being completed in the 1980s. Along with all other nuclear power construction, export activity ceased after the 1986 Chernobyl accident.

Note that, in the 1970s, US manufacturers of nuclear reactors needed to compete for export contracts on two fronts: the technology, and the financial package that could support the initiative. The need to provide financing places the US manufacturers at a serious disadvantage with respect to countries like Russia and China, which had state support.

American manufacturers competing for foreign contracts have access to support from a government agency: the Export–Import Bank of the United States (EXIM). While this is a more arm's-length relationship in support of financial backing than experienced by counterparts in Russia or China, it is an agency that can facilitate contract negotiations. EXIM is an independent Executive Branch agency of the American government with a mission of supporting the American economy through the facilitation of the export of US goods and services. Because it is backed by the full faith and credit of the United States,

EXIM assumes credit and country risks that the private sector is unable or unwilling to accept. The agency's charter requires that all transactions it authorizes demonstrate a reasonable assurance of repayment. EXIM consistently maintains a low default rate and closely monitors credit and other risks in its portfolio.

Since 2022 there has been an upsurge of US Government support and advocacy for US nuclear exports. This has included promotion of US nuclear marketing in Eastern Europe and a major commitment by EXIM to support sales of SMRs. They are ready to provide a strong competitive set of export finance tools that can help American SMR designs compete globally and help potential clients secure large-scale, flexible financing to turn projects into reality.

In line with this new program EXIM issued a $3 billion support-of-financing letter in April 2023 for deployment of a GE–Hitachi BWRX-300 reactor in Poland. GE–Hitachi is actively marketing the BWRX internationally with its first implementation moving forward in Canada.

NuScale has been extremely active in promoting SMR technology in the Czech Republic, Poland, Romania, and Türkiye. Despite problems NuScale has encountered in the United States, their foreign outreach is being advanced with vigor.

Great Britain is planning to build SMRs as part of its plan to reduce carbon emissions. In addition to the participation of Rolls-Royce in the competition for these sales, exports from other countries are being considered. In a review of potential SMR providers a short-list was prepared in October 2023, which included: Rolls-Royce, EDF, GE–Hitachi, Holtec Britain, NuScale, and Westinghouse. The US company TerraPower, founded by Bill Gates, was not included presumably because it depends on the use of 20% enriched uranium that, as of early 2024, is only available from Russia.

With support from the US Department of State, Westinghouse is making substantial progress with the export of full-scale 1000-MW class reactors. Agreements have been signed by Westinghouse for sale of AP1000 reactors in Poland and in Bulgaria with likely participation in nuclear power advancement in Türkiye.

Summary

There is a robust and expanding marketplace for nuclear reactor technology. While Russia has established a dominant position in this arena, China and the United States are increasingly competitive, with South Korea, France, Canada, and the United Kingdom becoming active players. It must be recognized that the competition is not purely economic, as relationships are used to leverage international influence, as illustrated by China's BRI. While the development of UN and other multilateral efforts to effectively manage global warming clearly influence a steady trend toward carbon zero, the competition evolving in the marketplace will certainly have at least equal impact.

Further Reading

Jessica Loverling and Havard Halland, "Russia's Nuclear Power Hegemony; The Wes is Dependent on Moscow for More Than Just Gas and Oil," *Foreign Affairs*, June 8, 2022
Discussion of the dominance of Russia in the nuclear export marketplace.
Matt Bowen. *Global Future of Nuclear Energy*, Atlantic Council Global Energy Center, August 25, 2023.
A research report on nuclear export programs.
Nevina Schepers 2019. *Russia's Nuclear Energy Exports: Status, Prospects, and Implications*, Stockholm International Peace Research Institute.
Zongyuan Zoe Liu, "Renewing America's Leadership in the Global Civil Nuclear Energy Market," *Council on Foreign Relations* [Blog], June 22, 2022 https://www.cfr.org/blog/renewing-americas-leadership-global-civil-nuclear-energy-market
Account of new initiatives in the 2020s by the United States to reassert its presence in the nuclear marketplace.

25

Meeting the Global Warming Challenge

Nuclear Contribution to Net Zero

The 28th United Nations Climate Change Conference (COP28) was a historic event for nuclear energy when it was formally specified as one of the solutions to climate change in the First Global Stocktake of progress toward meeting the goals of the Paris Agreement.

– Nuclear Energy Agency[1]

Climate Change is Real

The evidence that human-induced climate change is causing ever-increasing damage to the environment is overwhelming. Two authoritative references that support this threat to worldwide stability are analyses of the Intergovernmental Panel on Climate Change (IPCC), a scientific body established by the United Nations, and the World Meteorological Organization (WMO).

The IPCC Climate Change 2023 Synthesis Report contains many detailed results of studies and observations. These studies meet the highest standards of objectivity and scientific rigor. Among the noteworthy conclusions stated with confidence are:

- Human activities, principally through emissions of greenhouse gases, have unequivocally caused global warming, with global surface temperature reaching 1.1 °C above 1850–1900 in 2011-2020.
- Widespread and rapid changes in the atmosphere, ocean, cryosphere, and biosphere have occurred. Human-caused climate change is already affecting many weather and climate extremes in every region across the globe.
- In all regions increases in extreme heat events have resulted in increased human mortality and morbidity.

---

[1] Nuclear Energy Agency, "COP28 Recognises the Critical Role of Nuclear Energy for Reducing the Effects of Climate Change," December 23, 2023. https://www.oecd-nea.org/jcms/pl_89153/cop28-recognises-the-critical-role-of-nuclear-energy-for-reducing-the-effects-of-climate-change, accessed January 30, 2025.

Nuclear Energy. Edward A. Friedman, Oxford University Press. © Edward A. Friedman (2025). DOI: 10.1093/9780198925811.003.0025

Clearly, climate change is a well-established reality and if vigorous programs to limit greenhouse gases are not implemented, continued global temperature increases will bring increasingly severe consequences. Given that every country of the world emits greenhouse gases and that the consequences of these emissions affect all people of the globe, solutions require international participation.

As we look to the future, we confront the question: "How will society eliminate greenhouse gasses by 2050?" This date of 2050 is a pragmatic choice that envisions saving the world from the direst consequences of global warming.

With electricity and heat production responsible for a third of global carbon dioxide-equivalent emissions, and with many industrial processes able to be powered by electricity, the value of non-carbon-emitting sources of energy is immense. Given that there is considerable uncertainty about the rate of increase in average world temperatures, as well as the nature of these changes in climate, some combination of wind, solar, and other renewables together with nuclear power needs to be implemented. International organizations, agencies, think tanks, research centers, and pundits have proposed various scenarios for interventions that could protect the world from potential climate-induced devastation. Although there is little doubt that the 2020s witnessed the emergence of a nuclear renaissance, the impact of this development, when seen in the context of the world achieving net-zero by 2050, remains challenging.

It is beyond the scope of this book to identify an optimum scenario to reduce the dangers of global warming, which identifies the contributions to carbon reduction from the various non-carbon emitting sources. However, it can be said with certainty that nuclear energy will play a significant role.

Tackling this issue requires having a clear picture of what needs to be done and how it might be accomplished. There are many facets to the challenge of achieving "net-zero" and many possible responses to those challenges. What follows is an attempt to identify those distinct challenges and to articulate some of the most promising responses, which could include the use of nuclear technology.

The limitations of wind and solar energy will require incorporation of nuclear energy in a comprehensive worldwide solution to "net-zero" challenge. This is because wind and solar power are not available 24/7 and the use of batteries to balance the load remains problematic. Furthermore, wind and solar power generation require large areas for deployment of apparatus. The placement of wind and solar generating plants at locations that are distant from densely populated urban areas is problematic because few locations have power grids capable of transporting electrical energy long distances. It also must be recognized that there are regions of the world where installation of wind and/or solar energy plants is simply not practical. Weather conditions and remoteness are factors that can prevent development of wind and solar solutions.

What is often overlooked in discussions about solutions to global warming is that much of the industrial sector (discussed in the next section) that contributes 23% to carbon emissions uses high-temperature direct heat that cannot easily be provided by electricity. However, the use of nuclear energy to meet this need is evolving in a promising fashion.

The Five Sectors Requiring Greenhouse Gas Reduction

While the term, "net-zero," appears in nearly every discussion regarding the global warming challenge, it is most frequently mentioned in the context of carbon dioxide emissions that result from the production of electricity. While reduction of carbon emissions in electricity production is of prime concern, it is only part of the picture. In the United States, greenhouse gas emissions originate in five sectors of the economy (as of 2021): Electric power 25%, transportation 28%, industry 23%, commercial and residential 14%, and agriculture 10%.

1. **Electric Power (25%):** This sector involves the generation, transmission, and distribution of electricity. Carbon dioxide, methane, and nitrous oxide are released during the combustion of coal, oil, and natural gas, whereas negligible quantities of these gases are emitted by nuclear, solar, wind, operational hydropower facilities, and other renewable energy sources.

2. **Transportation (28%):** This sector produces carbon emissions from fossil fuels and is amenable to significant carbon reductions through the replacement of fossil fuel propulsion systems with energy provided by electric- or hydrogen-powered batteries. The most progress in this transition is occurring in automobiles, buses, and trucks, and the least progress is in aviation, which globally accounts for 2.8% of emissions.

There has been considerable momentum for the use of electric-powered automobiles as well as the introduction of hydrogen fuel cell automobiles. Electric buses and light trucks are in use and heavy-duty trucks are likely to be on the road when more refueling stations become available.

3. **Industrial (23%):** This sector's emissions are produced by burning fuel for production of products, including plastics, pharmaceuticals, metals, and cement, with added emissions often resulting from smelting or other chemical processes directly. The four largest sources of greenhouse gases in this sector originate in the production of cement, steel, ammonia, and ethylene. This sector also includes power for industrial buildings and machinery. Cement and concrete production generate about 9% of all carbon dioxide emissions. Concrete is the second most widely used substance on earth with water being the first. Close behind concrete in carbon emissions is the iron and steel industry, which accounts for about 7% of global greenhouse gas emissions. Production of other metals, such as aluminum, also uses high-temperature direct heat. Ammonia production contributes approximately 2% of all carbon dioxide emissions worldwide. Around 80% of ammonia that is produced industrially is used to make nitrogen-based fertilizer that enables around half of the world's food production. It is also used as a building block to make pharmaceuticals and many commercial cleaning products. This widely

used chemical consists of three hydrogen atoms and one nitrogen. It was first manufactured in 1909 by two German chemists who made it from water that contributed the hydrogen which combined with nitrogen that is present in ambient air. Germany used ammonia during the First World War to produce explosives. Another chemical that is responsible for large carbon emissions is ethylene, which is perhaps the most important petrochemical raw material. It is a component in countless chemical reactions, and an essential ingredient in various plastics and detergents among other chemical products. It is estimated that just under 1% of the world's carbon emissions originate in the production of ethylene.

4. **Commercial and Residential (14%)**: This sector includes energy consumed by homes and commercial businesses as well as energy required for managing waste, wastewater, and leaks from refrigerants in homes and businesses.

5. **Agricultural (10%)**: This sector contributes greenhouse gases from the use of fertilizer, the digestion and excrement of livestock, as well as ground treatment and burning of crop residues. This 10% of the total greenhouse gas burden cannot be met with alternative power sources. Processes such as carbon capture will be required to mitigate the impact of this production of greenhouse gases.

Opportunities for Reducing Greenhouse Gases with Nuclear Energy

The primary use of nuclear energy to date has been to produce electricity. As a direct replacement for the baseload services of coal-fired power stations, nuclear power has considerably reduced the portion of electricity derived from fossil fuels worldwide, with the potential for even further reductions. Several countries and jurisdictions have largely decarbonized their electricity supply using nuclear power in combination with other sources, predominantly hydropower. These successes can be repeated today with Generation-III+ reactors and, in coming years, with Generation-IV reactors.

Generation-III+ reactors are light-water reactors that incorporate passive safety features that do not require active controls or operator intervention to mitigate abnormal events. Instead, the laws of physics are relied on for the actions of gravity or natural convection to provide protection from overheating. By innovating proven reactor designs rather than leaving them, they can navigate regulatory approvals more quickly and offer a shorter path to commercialization than Generation-IV reactors.

A Generation-III+ design developed by Westinghouse and improved upon by manufacturers in China and South Korea is the AP1000. This reactor was designed to produce 1117 megawatts (MW) of electrical energy using a somewhat simplified design that incorporated elements of modular construction. Westinghouse has had a difficult time in constructing the AP1000 in the United States. Work began on developing this reactor in Georgia in 2013. Delays and difficulties led not only to the cancellation of plans for units that were to be built in South Carolina but also to the 2017 Westinghouse bankruptcy. Two AP1000s became operational in Georgia in 2024 at the Vogtle

Electric Generating Plant. These reactors came online after long delays and hugely over budget. This experience of Westinghouse with AP1000s is frequently cited as reason to reject nuclear reactor technology. However, others have improved on this design to build successful power plants in a timely, cost-effective implementation.

China's nuclear power corporation devoted 12 years to develop an enlarged and improved version of the AP1000 known as the CAP1400 (or Guo He One). This design became operational in 2020 and is being promoted for the export market.

The Korea Electric Power Corporation (KEPCO) began construction of a power plant with four APR1400 units for operations in the United Arab Emirates. As of 2024, all four units, each producing 1400 MW electrical, are operational. KEPCO is actively pursuing additional export opportunities for the APR1400. The nuclear industry has not shown itself to be entirely cooperative, but rather competitive when it comes to international reactor sales. In one recent example, Westinghouse, the intellectual property owner of the System 80 reactor on which the APR1400 is based, filed a lawsuit seeking to prevent KEPCO from marketing the APR1400 in Poland. It claims that, by sending Poland a bid, KEPCO violated US export control laws. However, this case was dismissed in September 2023 by a US district court, although a final ruling by the arbitration panel is not expected until late 2025.

Russia began construction of VVER light-water pressurized reactors in the late 1960s. There has been steady modification and improvement of this design with the implementation of a Generation-III+ model in 2018 that is now operational in Kursk, Russia. Four Generation-III+ reactors have been designated for use in Türkiye with expected operational dates around 2024. Rosatom, the Russian state nuclear corporation, projects a 40-month construction time for these advanced VVER reactors, which would be a breakthrough for reactor deployment.

Westinghouse, China, South Korea, and Russia are all engaged in the promotion of their Generation-III+ reactor designs for export. Canada, France, and India are pursuing development of competing designs. With the emergence of this proven fail-safe technology, there is ample opportunity for implementing programs to reduce carbon emissions in the electric power sector with nuclear power.

Evolving Momentum for Generation-IV Technology

While skeptics scoff at the potential realization of Generation-IV nuclear technology by 2030, it must be remembered that Generation-IV nuclear technology has evolved systematically over a $\geq$ 30-year history of unprecedented technological planning and development. The success of this international collaborative initiative could not have been anticipated at its origin in the early 2000s following the accident-induced hiatus in nuclear construction. This extraordinary record of achievement, if anything, should underpin optimism that the programs objectives will come to fruition.

As Generation-IV technology becomes operative there will be many additional opportunities to provide electricity without carbon emissions. The development of sodium-cooled, lead-cooled, and molten salt-cooled reactor technology also provides opportunities to use nuclear energy in locations that do not have ocean- or river-based sources of coolant water. Although Generation-IV reactors are immature technologies in comparison with Generation III+, significant interest in their potential applications has driven funding and talent toward their demonstration and commercialization across the globe.

The elimination of greenhouse gases produced in the transportation sector becomes possible as motor vehicle propulsion systems are increasingly powered by electricity rather than fossil fuels. The use of nuclear energy to provide this electricity for the transportation sector will be especially important in areas that cannot support energy generation by wind or solar systems. This might be the case where space is not available or where the electric grid cannot easily link the location where the energy is generated to the location where it is needed.

The realization of the first Generation-IV modular high-temperature gas-cooled reactor (HTGR) was realized in China in 2023. The ability of China's nuclear industry to actualize a reactor that can drive a 210-MW steam turbine opens multiple opportunities to reduce greenhouse gas emissions. HTGRs can potentially be used in the industrial sector where they can provide process heat directly that is free of carbon emissions.

The promise for successful realization of Generation-IV technology was significantly enhanced with the groundbreaking of the TerraPower sodium-cooled Natrium reactor (see Chapter 19). On June 10, 2024, when Bill Gates's company TerraPower broke ground for the construction of the Natrium fast-neutron, liquid sodium-cooled reactor (SFR) that uses an associated liquid-salt energy storage system. This fail-safe, small modular reactor (SMR) will be the first model in the world for cost-effective replacement of a coal-fired power plant with a nuclear reactor. This facility is projected to become operational by 2030. Even at this early stage of implementation, TerraPower is engaged in planning to emulate this development in Kemmerer, Wyoming at multiple locations in the US Pacific Northwest. Gates envisions hundreds of coal-to-nuclear transformations in ensuing years.

TerraPower has secured support for this project from the Federal Advanced Reactor Demonstration Project for up to $2 billion. Gates himself has provided well over a billion dollars of private capital for the Natrium reactor and has pledged to absorb the cost of budget overruns as construction moves forward. Given that there is significant successful experience with SFRs, that TerraPower has been a leader in using supercomputer simulations in the design of the Natrium reactor, and that sufficient capital is available for project implementation, the probability of successful realization of these project goals is quite high.

Potential Role for Hydrogen

A major application of nuclear energy that promises to lead to significant reductions in carbon emissions is in the production of hydrogen, which in turn has many carbon replacement applications. While the use of hydrogen to reduce carbon emissions has been promoted for many years, the cost of doing so has been prohibitive. With the availability of expanded nuclear energy on the horizon, the prospect of actualizing a "hydrogen economy" is within closer reach.

Hydrogen, the most abundant element in the universe, was identified by Robert Boyle in 1671 as a flammable gas emitted from iron filings interacting with dilute acid. In 1800, William Nicholson produced hydrogen gas by passing an electric current through water—a process known as electrolysis. Electricity for electrolysis and energy for a more recent thermochemical process for production of hydrogen can be provided via a nuclear cogeneration process from large nuclear reactors or with dedicated smaller reactors operating at high temperature. These applications of nuclear energy provide an avenue for the wide scale use of hydrogen in processes that do not release greenhouse gases.

Hydrogen fuel can produce the intense heat required for industrial production of steel, cement, glass, and many chemicals and pharmaceuticals. Its use in producing both ammonia and ethylene are among the many applications that are possible.

Its use in the transportation sector has the potential to compete with electric vehicles. Hydrogen-propelled cars have been introduced by Toyota, Hyundai, and Honda. Sales of these vehicles in California where they are marketed have been stymied by the scarcity of refueling stations. Market forces will need to play out in coming years to see whether the transportation sector is dominated by electric vehicles or hydrogen vehicles.

While a transition to the use of hydrogen in aviation is particularly challenging, there are serious initiatives to pursue that application. The Clean Aviation Joint Undertaking is the European Union's leading research and innovation program for transforming aviation toward a sustainable and climate neutral future. Contributing to this program is the Siemens Digital Industries Software development program that is pursuing two hydrogen-based solutions for powering long-haul carbon-free aviation. One is by means of liquid hydrogen-powered jet engines and the other is to employ hydrogen fuel cells that combine hydrogen and oxygen to produce electricity that can be used to power electric motors.

Mobilizing to Implement a "Net-Zero" Program

This book has attempted to provide a context for understanding the nature of nuclear energy technology and its development. How this development will evolve depends on many variables, including: growth in electricity demand; the strength of climate policies; levels of international cooperation; the receptivity of the public to nuclear; time to process licenses; availability of financing and incentives; engineering solution for Generation-IV reactors; and the time frame for creating and testing new designs. Given the track record for use of Generation-III+ designs, a time of approximately eight years needs to be expected between breaking ground for a new reactor and its connection to the grid.

Several new designs are expected to become operative by 2030 or soon thereafter. Scenarios for achieving net-zero are expected to play out between 2030 and 2050. While none of these scenarios may protect the world from experiencing serious consequences precipitated by global warming, it is important for an informed citizenry to have perspective on the implications of public policy decisions relating to these developments. It is in that spirit that this book has presented information on these topics.

In a world in which competing interests and ideologies motivate the behavior of nations, it is challenging (to say the least) to implement policies and programs that span the globe. However, there are two international organizations that can grapple with the integration of nuclear energy policy into the quest for solutions to global warming. They are the International Atomic Energy Agency (IAEA) and the Nuclear Energy Agency (NEA) of the Organization for Economic Co-Operation and Development (OECD). With 178 member states, the IAEA addresses policy issues as well as acts in a governance role on multiple issues concerning nuclear program activity. The IAEA was founded in response to the need to have an international agency that coordinated policies and programs relating to nuclear weapons. It is constituted as an autonomous organization within the United Nations system. The NEA, with 38 members, promotes programs and policies that have an economic development component. It is an integral part of the OECD, which mainly addresses economic issues. Both agencies issue updates on the quest for net-zero along with recommended action plans.

These international agencies, as well as special committees of the United Nations, have issued analyses and calls to action that have common elements. The most succinct and ambitious statements came from the NEA and the OECD. In support of an analysis done by the IPCC, the NEA calls for a threefold increase in nuclear energy production by 2050 compared with the level that existed in 2020. Given a base in 2020 of 444 power reactors providing 394 gigawatts (GW) of electrical energy, an increase of nuclear energy needed to sustain a net-zero environment in 2050 would require 1160 GW of electrical energy. This figure of 1160 is almost exactly three times the 2020 base number of 394. Using that same multiplying factor for the number of nuclear reactors yields a figure of 1332 which is a total that exceeds the current 2020 number of 444 by 778. However, one must consider that many of the existing reactors would exceed their lifetime limits and be closed by 2050. Roughly speaking there would be a need to construct about 900 1000-MW reactors during the period from 2030 to 2050, or about 45 per year. Given that if such a pursuit were undertaken by eight reactor-producing countries (Russia, China, United States, France, South Korea, India, United Kingdom, and Sweden) it becomes possible to envision such an agenda becoming a reality.

Critics deride the prospect of such a massive undertaking. The implementation of such an intensive construction program of nuclear reactors is unprecedented and seems unlikely given recent history. However, this author believes that the equally unprecedented devastation likely to be experienced due to the increasing impact of global warming will stimulate society to take drastic action—with crash programs of this sort being implemented.

At the 28th United Nations Climate Change Conference (COP28) this goal of tripling the current supply of nuclear energy by 2050 was endorsed by 22 countries. This declaration was announced by French President Emmanuel Macron at a ceremony on December 2, 2023, at which he referenced the NEA analysis. He endorsed an expansion of nuclear energy output by stating, "If you want to reconcile jobs creation, strategic autonomy and sovereignty, and low carbon emissions, there is nothing more sustainable and reliable than nuclear energy."

The Declaration to Triple Nuclear Energy was endorsed by nations across four continents: Bulgaria, Canada, Czech Republic, Finland, France, Ghana, Hungary, Japan, South Korea, Moldova, Mongolia, Morocco, Netherlands, Poland, Romania, Slovakia, Slovenia, Sweden, Ukraine, the United Arab Emirates, and the United States. The declaration also emphasized the need for high-level political engagement and innovative financing mechanisms to mobilize investments in nuclear power.

However, there have been several critics of this nuclear initiative who have voiced skepticism about its attainability. *The New York Times* quoted a representative in the oil industry, who stated that, " . . . the pledge was divorced from the reality of nuclear energy—that it was too costly and too slow."

The Potential for Small Nuclear Reactors

At COP28, NEA also launched an important initiative promoting the incorporation of SMR technology in the pursuit of Net Zero called Accelerating SMRs for Net Zero. This initiative captures the growing enthusiasm for the incorporation of SMRs in meeting the challenge of achieving net zero by 2050 that has emerged worldwide. The NEA initiative seeks to ". . . focus on establishing practical, solutions-oriented activities supported by a comprehensive plan of collaboration and knowledge exchange. The goal is to inform policy and investment decisions by governments, industry players and the financial sector."

A vote of confidence in SMR technology occurred in the fall of 2024 when two industrial giants entered into agreements to use SMRs to meet their existing needs. These corporations, who are deeply committed to the development of AI technology, with its enormous energy requirements, are seeking SMR solutions to meet those needs. Amazon has contracted with Energy Northwest, while Google has entered into an agreement with Kairos Power, each for multiple SMR units that will be dedicated to powering AI facilities. Another nuclear initiative in support of Intelligent AI is the deal reached by Microsoft to restart operations at Three Mile Island that was also implemented in the fall of 2024.

In reporting on this Microsoft initiative, on November 11, 2024 *The New York Times* noted that "Goldman Sachs projects that U.S. power demand will grow more than 2% a year on average through the end of the decade, with data centers consuming around 8% of the country's electricity by 2030, up from roughly 3% now."

Immediate priorities for this initiative must confront several areas of concern.

Legal, Logistic, and Regulatory Challenges

Given the many uncertainties surrounding civilian technology that emerged from weapons research and development origins, it is unsurprising that commercial entities were reluctant to engage in reactor development and use in the 1950s. High among these issues was that of financial liability in the event of accidents. To address this barrier to private participation in non-military use of nuclear energy, the US Congress enacted the Price–Anderson Nuclear Industries Indemnity Act (PAA) in 1957. The PAA provides compensation for the nuclear industry for liability claims arising from nuclear incidents while providing coverage for the public. In March 2024, Congress extended this act to the end of 2065 with a provision tying liability to inflation and providing coverage for events outside the United States up to $2 billion.

Skeptics have voiced concerns that even the most promising plans for a nuclear energy breakthrough would not survive the bureaucratic hurdles that have prevented a full-scale nuclear reactor design being approved in the United States since 2012. Multiple concerns regarding impediments for nuclear reactor development were addressed with the signing into law of the Advanced Nuclear for Clean Energy (ADVANCE) Act on July 11, 2024.

This legislation seeks to modernize the nation's approach to licensing new reactor technologies through:

1. Reduced costs for licensing advanced nuclear technologies.
2. Incentives for deploying next-generation reactors.
3. Streamlining the licensing process for advanced nuclear technologies.
4. Promoting the development of SMRs that are small enough to fit on a tractor-trailer and be deployed at remote sites.
5. Expanding the development of nuclear reactor fuel.
6. Funding reuse of coal-powered plant sites for development of nuclear facilities.

This legislation received overwhelming bipartisan support. Nuclear energy development is one of the only consequential issues that receives enthusiastic endorsement from both US political parties.

Technological Challenges

The NEA Small Modular Reactor Dashboard, published in November 2023, identified 56 SMRs in various stages of active development. While designation of small nuclear reactors refers to reactors with up to an output energy of 300 MW electric, NEA includes the Rolls-Royce SMR designed to provide 470 MW electric. Other high energy-producing reactors included in the NEA compilation are the lead-cooled fast reactor (LFR), which has a designated output energy of 300–400 GW electric, being developed jointly by Westinghouse with Italian company Ansaldo Nucleare; the sodium-cooled Natrium reactor with output energy of 345 MW Electric being developed jointly by TerraPower and GE–Hitachi; the BWR GE–Hitachi BWRX-300 with a designated output energy of 300 MW electric; and the Integral Molten Salt Reactor with a designated output of about 300 MW electric being developed by Terrestrial Energy.

At the low end of energy production there are many reactors in development or being planned. Three that are noteworthy have been awarded a total of $3.9 million by the US Department of Energy for development at an Idaho National Laboratory test bed. They are the 5-MW electric eVinci reactor from Westinghouse, the Ultra Safe Nuclear Corporation's Pylon reactor with output power of 1.5–5 MW electric, and the Radiant Industries Kaleidos reactor with output power of 1.2 MW electric.

NEA in its position statement on SMRs emphasizes two major advantages that have gained the enthusiastic attention of so many development groups. One advantage is their financial viability, and a second advantage is their potential use in high-temperature applications.

A SMR's lower cost makes financing of construction more attainable. Further, its small size means that construction times can be much shorter, thus allowing more rapid return on investment. In cases where SMRs also meet design criteria for Generation-IV nuclear technology, factory fabrication provides greatly enhanced development opportunities. The 2024 partnership established between Korea Hydro and Nuclear Power with ARC Clean Energy and New Brunswick Power was in direct response to the emerging market in Southeast Asia for SMRs. The Philippines, Indonesia, Malaysia, and Singapore have all indicated interest in SMRs. The megacities of Southeast Asia are hungry for carbon emission-free power to support their growth.

Market Opportunities

The potential for high-temperature applications places nuclear reactor technology in a category that cannot be matched by any other alternative carbon-free energy source. The existing fleet of water-cooled reactors are unable to operate at temperatures exceeding 1292 °F (700 °C), while some Generation-IV SMRs that use gas cooling or molten salt cooling can attain output temperatures of up to 1832 °F (1000 °C).

High-temperature applications, which account for more than 20% of carbon emissions, include steelmaking and other metals processing, petrochemical processing, and thermochemical hydrogen production. HTGRs are the only heat source that can provide carbon-free energy for these and related applications.

Of these uses for HTGRs the production of hydrogen stands out as a strategic opportunity since hydrogen can be used to improve multiple situations in which carbon emissions are produced. Hydrogen does not occur naturally and must be produced via industrial processing. This has been done in chemical processing of natural gas and of gasified coal. Both cases contribute to carbon emissions. In contrast, using direct heat from nuclear reactors can be used to decompose water for hydrogen production without adding carbon into the environment.

Hydrogen has the potential of being used in a wide array of applications, including in the transportation sector, where hydrogen-powered vehicles are currently in limited use. In these cases, hydrogen batteries employ ionized hydrogen to combine with ionized oxygen (from the air) to produce an electric current with benign water as the exhaust.

Desalination is another application of HTGRs. Worldwide growth in the demand for potable water is likely to grow substantially by 2050. The ability of producing potable water from sea water using nuclear reactors that are configured to co-generate power and heat would be an economic use of HTGRs.

Reactor designs that can attain these desirable high temperatures are either configured with molten salt or are gas cooled using helium gas. An HTGR achieved operational status for 50 days in Japan in 2004. This test reactor's outlet temperature reached 1742 °F (950 °C), which was a world record. This reactor had a power level of only about 10 MW electric. The reactor has been offline since 2011, but plans are being developed to use the facility to produce hydrogen.

In December 2022 in the Shandong province of China, successful operation of pebble-bed HTGRs was attained. This power plant is configured with two small reactors that drive a single 210-MW electric turbine. This was the world's first demonstration of a continuously operating reactor that can produce high-temperature heat for industrial processes. HTGRs can not only be used directly for high-temperature industrial processes but also can be effective in the generation of hydrogen, which can be stored and shipped for use in implementing industrial processes requiring temperatures in excess of 2732 °F (1500 °C), such as steel and concrete production. Decarbonization of such high-temperature processes is especially challenging for wind, solar, and water-cooled nuclear power plants.

A second reactor technology that can provide high-temperature output is the Molten Salt Advanced High-temperature Reactor (AHTR). This design is receiving considerable attention due to its use of molten fluoride salts with boiling points that are near 2552 °F (1400 °C) that allow operations at high temperature and atmospheric pressure. Such a reactor would have the capability of producing hydrogen in processes at 1832 °F (1000 °C) economically and with a high degree of safety. There are several salts that can be used, with all of them having superior heat transfer characteristics compared with helium. A major challenge in the development of reactors that can operate in the range of 1832 °F (1000 °C) is the availability of materials for construction of the reactor that can maintain their strength and integrity at such a high temperature. Advocates for AHTR designs believe their use would yield significant economic advantages in electricity and hydrogen production.

Fuel Availability

With the dominant position occupied by Russia in nuclear fuel production, the conflict in Ukraine has precipitated disruptions in fuel availability. The United States and Europe have responded with initiatives to provide alternative sources of enriched uranium fuel that will meet the growing demand for fuel, albeit with delays of two to three years in production schedules. In the United States the Emergency National Security Supplemental Appropriations Act of 2024 provides funding for conventional and advanced uranium enrichment capabilities with an allocation of $2.7 billion. Substantial funding has been earmarked specifically to produce high-assay low-enriched uranium (HALEU) fuel that is required to operate the TerraPower Natrium reactor and other designs that require 20% uranium enrichment levels. While meeting energy production requirements, the expanding use of HALEU raises weapons proliferation concerns since uranium at that level of enrichment can be used in bomb construction. Widespread use of HALEU-fueled reactors will require scrupulous attention to fuel management to protect against weapons proliferation.

Summary

The emerging array of nuclear reactors—both large and small—holds the promise of providing significant reductions in greenhouse gas emissions in coming years. Nuclear, together with wind and solar energy, can eliminate more than 90% of carbon-emitting energy sources. The gap that these energy sources cannot fill will need to be addressed by carbon capture and carbon sequestration strategies.

Ideally, one would like to see a world government that could organize all the countries of the globe into a coherent effort to eliminate global warming. Given the reality of a fragmented world there still seems to be a way forward through competition among the nations that are clean energy producers. While such a competitive world is evolving, creative solutions will be needed to finance the implementation of needed technologies in resource-poor countries.

We see that the nuclear renaissance is evolving on many fronts. The COP28 pledge by 22 countries to triple the supply of nuclear energy by 2050 is on a path to realization of this goal. Multiple instances of initiatives in individual countries to meet this goal can be cited. Notable is the funding being allocated in the United States, where Congress has seen political parties that are at odds with each other coming together in an astonishing show of unanimity to pass legislation in support of an expanded nuclear future. In particular, Congress is supporting Advanced Reactor Demonstration Projects which in effect are public–private partnerships that enable private developers to achieve success in building first of a kind promising designs. These expensive demonstration reactors are difficult to fund based solely on marketplace financial projections. Examples include a government commitment to Kairos Power of up to $303 million for fabrication of a high-temperature fluoride salt-cooled reactor to be built in Oak Ridge, Tennessee. Similar support is being provided for the Terrapower Natrium Reactor planned to replace a coal-powered power plant in Kemmerer, Wyoming, with a SFR and the X-Energy Xe-100 HTGR planned for a Dow Chemical facility in Seadrift, Texas. If successful, these three Generation-IV designs will be operative in the United States before 2030 and set the stage for industrial replication to advance the COP28 goal. In addition to funding, Congress has passed legislation to promote streamlining of licensing for next-generation nuclear technology.

In Europe the nuclear renaissance is being actively promoted by the leadership of France, where the President Macron has been campaigning for the European Union to take a more proactive nuclear position. At COP28, Macron declared, "Nuclear energy is back!" He further stated that, " . . . we need the World Bank, international financial institutions [and] multilateral development banks to include nuclear energy into their energy-lending policies," https://www.euractiv.com/section/politics/news/macron-at-cop28-nuclear-energy-is-back.

The United Kingdom is implementing concrete plans to not only meet but to exceed COP28 goals outlined in a roadmap issued in January 2024 that seeks a fourfold increase in the use of nuclear power by 2050. In addition to expanding its fleet of nuclear reactors, Britain plans to become the first EU country to launch a high-tech fuel program with the goal of eliminating dependence on Russian fuel production.

Significant progress is also moving forward in India, where start-up was announced in 2024 of its first fast breeder reactor (FBR) in Tamil Nadu in which a mix of uranium and plutonium are being used with a blanket of uranium and thorium. This breeder's output of plutonium and uranium-233 will then be used as fuel in advanced heavy-water reactors. India will thus become the second country in the world, after Russia, to operate a commercial FBR. This long-term strategic approach will allow India to expand its nuclear energy production while making use of its vast reserves of thorium.

In August 2024, plans were also announced in India to build 40–50 SMRs (see Chapter 10) in factories. When announcing that initiative, managing director and CEO of Tata Consulting Engineers Amit Sharma stated, "To be honest, the only viable long-term solution for net zero is nuclear, I think nuclear is the bet; globally, everybody recognizes it," https://economictimes.indiatimes.com/industry/energy/power/plan-to-deploy-40-50-small-modular-nuclear-reactors-tata-consulting-engineers-ceo-amit-sharma/articleshow/112776277.cms?from=mdr.

The reality of a nuclear renaissance is evident in other countries where a retreat from nuclear energy following the events at Fukushima is now witnessing a turnaround. Prominent in this category are both Japan and South Korea, where a phase-out of nuclear energy was in progress and is now showing signs of being reversed.

While not signatories to the COP28 Nuclear Declaration, both China and Russia continue to pursue dynamic nuclear energy development programs and to seek to maintain leadership roles in the nuclear export marketplace. Major advances for each country have recently been manifested with Russia making progress in construction of a 300-MW electric Generation-IV lead-cooled fast reactor that is expected to become operational in 2026. This Generation-IV reactor will join the Generation-IV 800-MW electric SFR that has been online since 2023.

China is also pursuing development of several Generation-IV nuclear reactor designs with recent success in placing a HTGR online. Success with floating nuclear reactors and multiple designs for Generation-IV reactors promises to maintain both Russia and China as dynamic players in the export market. These safe next-generation designs that they will be promoting will facilitate acceptance in emerging markets such as Africa, Malaysia, and Indonesia.

Note that as solutions to the carbon challenge evolve, the challenge itself is becoming more severe, as the world population increases and concomitant energy needs increase with it. The current world population of more than 8 billion is projected to grow to

9.7 billion by 2050 according to UN agency projections. As a greater fraction of this larger population, especially in China, India, and Africa, engages in increased energy consumption, the need for increased amounts of carbon-free energy becomes ever more urgent. This population-driven need is simultaneously being exacerbated by energy-hungry applications that are emerging from expanded use of electric vehicles and the huge energy requirements of AI and bitcom technologies. If factory production of SMRs becomes a reality that application will certainly be employed in meeting this burgeoning clean energy need.

This author believes that as the climate crisis becomes more threatening, which will almost certainly happen, that the looming dire consequences will trigger new collaborative initiatives. As programs evolve, citizens not trained in science and technology will need to participate in the planning and implementation of constructive policies. It is hoped that this book will provide useful perspective in meeting those challenges.

Further Reading

Bill Gates, *How to Avoid a Climate Disaster: The Solutions We Have and the Breakthroughs We Need*, Alfred A. Knopf, 2021.
A comprehensive overview of policy initiatives that could lead to net zero. Gates, who owns the nuclear energy company TerraPower, includes nuclear energy in the mix.
International Atomic Energy Agency, *Nuclear Energy in Mitigation Pathways to Net Zero*, IAEA, 2023.
This IAEA publication addresses the contribution of nuclear energy in global climate change mitigation pathways as set forth by the Intergovernmental Panel on Climate Change (IPCC) Sixth Assessment Report and the International Energy Agency World Energy Outlook.
Joel Guidez, *Fast Reactors: A Solution to Fight Against Global Warming*, Elsevier Academic, 2022.
Reviews status of fast liquid-sodium and molten-salt reactor technology with a focus on ecology and sustainability benefits. The book covers safety aspects, short-life waste management and biodiversity preservation.
Rafael Mariano Grossi, "Nuclear Must Be Part of the Solution: Reinforcing the Bargain That Strengthens Security While Expanding Peaceful Use," *Foreign Affairs*, July 18, 2024. https://www.foreignaffairs.com/united-states/nuclear-must-be-part-solution
Grossi, the Director General of the International Atomic Energy Agency, explores the three interrelated policy areas: nuclear disarmament, nonproliferation of nuclear weapons and the peaceful uses of nuclear science and technology.
Suriya Jayanti, "Nuclear Power Is the Only Solution," *Time*, December 4, 2023.
Commentary on the announcement at COP28 with pledges from many countries to triple nuclear power use by 2050.

Assignments and Enrichment

A list of suggested assignments is provided to facilitate the use of this book as a college-level text. These assignments are largely focused on topics that enrich the content of the course beyond the material contained in the book.

1. Read *The Manhattan Project—Making the Atomic Bomb*. https://www.osti.gov/servlets/purl/10186004

Write 400–500 words on how the success of the Manhattan Project was enabled by the industrial resources (experience, capacity, and diversity) of the United States, with emphasis on the work at Hanford and the Clinton Engineering Works (Oak Ridge).

2. Read: "Nuclear Power Reactors" up to and including the section on CANDU reactors on the website of the World Nuclear Association.
    https://world-nuclear.org/information-library/nuclear-fuel-cycle/nuclear-power-reactors/nuclear-power-reactors.aspx

Write a commentary about these three designs: Pressurized water reactors, boiling water reactors, and CANDU reactors. Discuss the pros and cons for each design.

3. Write an essay on why Lise Meitner deserved to receive the Nobel Prize

4. Write an essay on why Enrico Fermi deserved to receive the Nobel Prize in 1938 even though he did not discover two new elements.

5. Research the energy crisis of the 1970s and write 400–500 words on the impact of that event on increased use of nuclear energy in North America and Europe.

6. Write an essay on Hyman Rickover's impact on nuclear energy policy in legislative and regulatory agencies of the American Government.

7. Write an essay for a high school student explaining the causes and lessons to be learned from the Chernobyl accident.

8. Write an essay on Alvin Weinberg's impact on nuclear energy policy in the public arena. Make note of the Alvin Weinberg Foundation that was active in the United Kingdome and the Thorium Energy Alliance that is active in the United States.

9. As nuclear energy emerged from being a government initiative to being part of the private energy establishment in the United States, there was discussion of having it continue as a nationalized industry. What were the arguments for and against that course of action?

10. Write an essay about the protests in the 1970s and 1980s that were mounted by the anti-nuclear movement against the construction of the Diablo Canyon Power Plant in California.

11. Write a report on the experience of "downwinders" who received damaging radioactive exposures resulting from proximity to the Hanford reactors that were built to produce plutonium.

12. Greenpeace has had an impact on nuclear energy policy in countries around the world, including New Zealand and Germany. Write an essay in which you analyze the role of Greenpeace in international discourse regarding nuclear energy, including commentary about its approach to this issue and how it has evolved.

13. Write an essay reviewing the history of nuclear energy policy and its status in Taiwan. Taiwan produces 90% of the world's advanced semiconductors and its dependence on imported fossil fuels for this industry's energy supply presents a compelling justification for nuclear energy. However, as of mid-2024, this issue remains unresolved.

14. The controversy and international ramifications of tritium release in water released from the Fukushima disaster site is ongoing. Write a report on the status of this issue and its ramifications for both the fishing industry in the Fukushima region and for Japan's relations with other countries in Asia.

15. Prepare a report on the deep-sea disposal of nuclear waste that occurred between 1946 and 1993. What are the pros and cons of this method of waste disposal?

16. There has been an almost cult-like support for the development of thorium-based reactor technology. Discuss the history of the Alvin Weinberg Foundation and the Thorium Energy Alliance and the prospects for wide-scale development of this technology.

17. What is your estimate regarding the potential for extensive use of ultra-small nuclear reactors with power output of less than 100 MW electric? Write a review in which you elaborate upon your answer.

18. Prepare an easy on the history, status, and outlook for the future for breeder reactors.

19. Prepare a summary of actions being taken in nuclear reactor-producing countries to revise the processing of licenses for next-generation nuclear reactors.

20. Assume that NetZero will be achieved by 2050 with some combination of renewables plus nuclear. What is your estimate of the percentage of the world's total energy that will be contributed by nuclear in that mix? Write an essay justifying your response to this question.

Further Reading

At the end of each chapter further reading titles are identified. These are mostly books and documents that provide in-depth material relating to the content of the book. If the book is used in support of a college course, these works provide opportunities to assign book reports and essays and as material for classroom discussions.

In response to rapid nuclear energy-related developments taking place, the author is publishing a monthly newsletter called *Nuclear Tomorrow*. The newsletter contains citations to articles that report on new nuclear energy initiatives regarding technology, policy, and construction. This newsletter can be located online by searching for: Substack Nuclear Tomorrow.

Index

For the benefit of digital users, indexed terms that span two pages (e.g., 52–53) may, on occasion, appear on only one of those pages.

Tables, figures, and boxes are indicated by an italic *t*, *f*, and *b* following the paragraph number.